Queensland
Review

Editor
Michael T. Davis *(Griffith University)*

Special Issue Editors
Iain McCalman *(Australian Catholic University)*
Kerrie Foxwell-Norton *(Griffith University)*

Associate Editors
James Forde *(Griffith University)*
Paul D. Williams *(Griffith University)*

Reviews Editor
Anthanasios Antonopoulos *(Griffith University)*

Editors Emeritus
Belinda McKay *(Griffith University)*
Kay Ferres *(Griffith University)*

Advisory Board
Patrick Buckridge *(Griffith University)*
Denis Cryle *(Central Queensland University)*
Marion Diamond *(University of Queensland)*
Mark Finnane *(Griffith University)*
Richard Fotheringam *(University of Queensland)*
Jessica Gildersleeve *(University of Southern Queensland)*
Claire Kennedy *(Griffith University)*
Jock Macleod *(Griffith University)*
Robert Mason *(University of Southern Queensland)*
Antonella Riem *(University of Udine)*
Yorick Smaal *(Griffith University)*
Cheryl Taylor *(James Cook University)*

VOLUME **28** NUMBER **2** **2021**

Queensland Review

VOLUME **28** NUMBER **2** **2021**

Special Issue
Between pride and despair
Stories of Queensland's Great Barrier Reef
and Wet Tropics rainforests

Published by Equinox Publishing Ltd.

UK: Office 415, The Workstation, 15 Paternoster Row, Sheffield, South Yorkshire S1 2BX
USA: ISD, 70 Enterprise Drive, Bristol, CT 06010

www.equinoxpub.com

First published 2022

@ Equinox Publishing Ltd. 2022

All rights reserved. No part of this publication may be reproduced or transmitted in any form or by any means, electronic or mechanical, including photocopying, recording or any information storage or retrieval system, without prior permission in writing from the publishers.

British Library Cataloguing-in-Publication Data
A calalogue record for this book is available from the British Library.

ISBN 978 1 80050 312 0 (paperback)

Queensland Review

VOLUME **28** NUMBER **2** **2021**

Special issue
Between pride and despair: Stories of Queensland's Great Barrier Reef and Wet Tropics rainforests

Editorial   77
Iain McCalman and Kerrie Foxwell-Norton

Reef and Rainforest Reflection
Beautiful shells and their connection to the Reef   80
Chrissy Grant

Article
Caring for colour: Multispecies aesthetics at the Great Barrier Reef   82
Killian Quigley

Reef and Rainforest Reflection
Coralations: Back to the breath   94
Irus Braverman

Article
Aquariums and human–animal relations at the Great Barrier Reef   98
Ann Elias

Reef and Rainforest Reflection
Basket case!   114
Carden C. Wallace

Article
Great Barrier Reef World Heritage: Nature in danger   118
Celmara Pocock

Reef and Rainforest Reflection
Sounds of silence   130
Diane Tarte

Article
Coal versus coral: Australian climate change politics sees the Great Barrier Reef in court   132
Claire Konkes, Cynthia Nixon, Libby Lester and Kathleen Williams

Reef and Rainforest Reflection

Urannah: The isolated home of rare species — 147
Peter McCallum

Article

Women of the Great Barrier Reef: Stories of gender and conservation — 150
Kerrie Foxwell-Norton, Deb Anderson and Anne M. Leitch

Reef and Rainforest Reflection

The Daintree Blockade: Making (radio) waves — 166
Bill Wilkie

Article

Drawing a line in the sand: Bioengineering as conservation in the face of extinction debt — 169
Josh Wodak

Reef and Rainforest Reflection

'Tourist fiction': Cassowaries in Mission Beach — 183
Leonard Andy

Epilogue

A reflection on the role of tourism within vulnerable biodiverse reef and rainforest regions – a case-study from Mission Beach and the Cassowary Coast — 187
Iain McCalman

Between pride and despair: Stories of Queensland's Great Barrier Reef and Wet Tropics Rainforests

Iain McCalman and Kerrie Foxwell-Norton
Iain.McCalman@acu.edu.au, k.foxwell@griffith.edu.au

The Great Barrier Reef and Wet Tropics Rainforests occupy a crucial but conflicted space in Queensland's history: once symbols of conservation triumph, they are fast becoming portents of ecological collapse. Until relatively recently, these reef and rainforest ecologies were icons of a rich natural and cultural heritage that has brought pride to Queensland and to the nation at large, while our First Nations communities can celebrate relationships to northern reefs and rainforests that span at least 60,000 years. The ancient Gondwana rainforests of the Wet Tropics match the biodiversity of the Great Barrier Reef, with both having been recognised for their 'outstanding universal value' in UNESCO World Heritage Listings. Stories have repeatedly celebrated their beauty and biodiversity, and their rich and complex associations with the local peoples and communities that live there.

Yet those inspiring stories of the Great Barrier Reef and Wet Tropics Rainforests have now taken a dire turn as the emergence of severe threats to the health of both ecosystems threatens death and demise. The damage caused by mass coral bleaching events, acidification and super-cyclones is paralleled by deforestation, fire and species extinctions within the rainforests. Rising land and sea temperatures are proving to be ecologically devastating for both these wondrous ecosystems and equally grim in their associated social, cultural and political ramifications. Yet we editors have been pleased to observe that the contributors to this special edition have still been able to find some sources of inspiration and hope within these calamitous outlooks.

In this special edition of *Queensland Review*, we invited the contributors to reflect on the Great Barrier Reef and/or the Wet Tropics, and their relationship with people and places nearby and further afield. Perhaps not surprisingly, we found that the urgency of the Barrier Reef's current crises has tilted the contributions in that direction. Nevertheless, the interwoven plight of the rainforests is explicit in some articles and implicit in others. We also chose to commission both scholarly analyses and personal reflections, entwining the professional with the personal to generate representative expressions of insight, expertise and experience. The authors of the personal reflections chose an object or a photo to stimulate their writing, again revealing diverse relations with these spectacular natures.

It is thus with a great sense of pride that we present the tapestry of these stories of the Great Barrier Reef and Wet Tropics Rainforest, where personal reflections are interwoven with scholarly articles to showcase a broader humanity of environmental care. Here scholarly analyses are ranged beside the private memories of individuals who have dedicated lifetimes to conserving and protecting these two natural wonders. We also hope this approach will dilute the arbitrary divisions between ourselves and nature, and capture something of the resulting richness and complexity of human and non-human relationships.

As is inevitable in any such a contemporary collection on these subjects, a mammoth question looms over our special edition — both spoken and unspoken. Anthropogenic climate change, as a consequence of the destabilising interference of humans in the natural systems of our planet, demands critical and sobering reflection on ourselves. Should we, as humans, nevertheless attempt to save the Reef and the rainforest by creating a 'good Anthropocene' of restorative anthropogenic geophysical and technological interventions? Are there other ways of figuring human relations with our Reef and rainforests that might better secure a future for ourselves and the nature we love? While some contributors have addressed these questions explicitly, others have left their answers unstated. We editors do not attempt any definitive answers. Probably like most our contributors, we feel that that the jury is still out, although these are questions that humanities, arts and social sciences scholars will continue to ask.

We bookend our volume with personal reflections from Chrissy Grant, an Aboriginal (Eastern Kuku Yalanji) and Torres Strait Islander Elder (Mualgal from Moa Island) and Djiru Traditional Owner and artist Leonard Andy. As First Nations people with deep connections to Far North Queensland 'Country', their reflections rightfully launch our issue and demonstrate the enduring power of their cultural science and practice, and how these have also sometimes become enmeshed with Western systems of knowledge and management. These contributors bring into sharper focus the timelessness and indivisibility of First Nations natural and cultural practice and heritage.

We pair a sparkling provocation from Killian Quigley to refigure a proper caring for coral colours that elevates the vital importance of aesthetics in our reckoning of Reef value with Irus Braverman's rich personal reflection on 'coralations' as a challenge to thinking that similarly privileges the sensory. Ann Elias's pairing of art and environmental history to explore the 'reef aquarium' lithographs of 1930s tourism links with Carden Wallace's reflection on her later encounters as a coral scientist wielding the baskets used to collect corals. In both Elias's and Wallace's work, the exploration of underwater coral reefs from below and above the water's surface highlights continuing tensions, as well as odd synergies, between tourism and science.

Celmara Pocock's critical examination of World Heritage brings into sharper focus the actual frailty of ostensibly powerful international protections enlisted to protect the Reef. World Heritage listing based on the division of humans from nature — and the power of humans to commandeer Reef health — is and perhaps always has been 'in danger'. Anthropogenic climate change where the motherlode of human impacts and nature meet is evidence enough of the inseparability of human and non-human ecologies. Diane Tarte, as a founder of the Australian Marine Conservation Society and a fifty-year veteran of Reef protection, reflects on the despair that follows climate change impacts on Reef health, but finds hope in what has already been achieved, and what might be possible with the political will to limit global warming.

Claire Konkes, Cynthia Nixon, Libby Lester and Kathleen Williams skilfully dissect the recent furore surrounding the Adani Mine proposal and the 2015 legal challenge from the Australian Conservation Foundation which catapults the Reef into the mire and politics of media, law and protest. The ongoing work of the Mackay Conservation Group — a key player in Adani protests — to protect inland catchments is similarly outlined in Peter McCallum's reflection on the ongoing challenges of connectedness between the Reef and places distant. Kerrie Foxwell-Norton, Deb Anderson and Anne Leitch remind us, through the story of one valiant woman scientist, that the Reef is a gendered space that needs to revise its history in ways that elevate the achievements of women. Similarly, from the margins, Bill Wilkie celebrates passionate conservation activists, telling their Daintree Blockade story through a reflection on the humble radio log. The final scholarly article in the issue is Josh Wodak's vivid interrogation of turtle conservation on Raine Island, where the question of why and when we should intervene in conservation is complex and rife with contradictions about our best human work to save crucial species in an era of grave ecological threats.

Iain McCalman's Epilogue lands the key themes of this issue in reflections on his ongoing engagement with Mission Beach and the Cassowary Coast communities. These communities tread the fine line between caring for economy and ecology, where navigating the terrain traversed by our authors is an everyday lived experience.

We would like to thank *Queensland Review* editor Dr Mike Davis for his carriage of this project and Susan Jarvis for her well-renowned editorial prowess. Thanks to Camille Page for her editorial assistance and special mention too of Ms Christine Howes and Dr Valerie Boll for supporting personal reflections during a difficult pandemic when travel was not possible for us. We congratulate our authors and thank them for their commitment to this project. Finally, to our readers, we hope you enjoy this special issue even half as much as we have enjoyed collating it. Like us, may you find hope and comfort in the extraordinary minds, innovation and creativity of our authors. If you leave with nothing else, let it be the sense that in stories told of the Great Barrier Reef and Wet Tropics Rainforests, we are far from finished yet.

Beautiful shells and their connection to the Reef

Chrissy Grant
Eastern Kuku Yalanji (Cape York) and Mualgal (Torres Strait Islands) Traditional Owner
chrissy@webone.com.au

Shells are beautiful! They are really ingenious in the way that they are made and the animals they house. The shells grow with the animal, from tiny little shells to a great big shell. An animal wasn't born that big, so the large shells have been there for years.

The shells are things of beauty. A lot of shells are very fragile, while some of them are resilient to being tossed around in the current and wild seas.

The Nautilus shell is really paper thin, for instance. However, if you knock it you could crack it and then that's it, you've damaged it. But if you pick up shells and take a good hard look at them, there's hope. Hope that there's continuation of marine life in there, but at the same time, when you don't see them anymore, there's a bit of despair as well — especially if the shells barely exist in the numbers that used to be there and that's a shame.

So, shells for me are something that are not only decorative, but they tell a story. For the big clam shells, they're only found in certain areas. We had one when we were growing up because my father was a pearl diver and he had the diving helmet and walked on the seabed picking up shells for the pearls.

He was pearl diving up in Torres Straits, and we had a great big clam shell at home, you could sit a little baby in there. We had lots of shells around our place. My uncle owned lugger boats that went from Cairns to the Torres Straits, and they had shells around their house too.

That's where my first fascination about shells began.

You can listen to them, which is a really peaceful thing to do, especially when you're away from the ocean. My grandkids would put the shells up to their ears and say, 'Oh, listen to that, Nan!' with my reply, 'Yeah, that's the ocean you can hear — isn't that amazing!'

Maybe there aren't as many shells around now as there used to be, at least not the different varieties. There used to be dozens of different kinds of shells, different sizes, but also the different types; you can hardly get a big clam shell nowadays which I think is illegal to remove them from the ocean — at least, I hope it is illegal — to collect them anymore. They are rarely seen now, even in documentaries or footage being taken of someone diving on the coral reef. We need to do all we can to save the giant clams along with other marine species now!

I don't know about the severe weather or the impacts of cyclones on shells and things. It may be that they've had some impact on the breeding of them or the placing of them, because those big clam shells wouldn't move around very much.

It's really important for all the land and sea Country to be as pure as it can be without any interference from humans. There's a plan to put power in over the Daintree River now, and once power is laid on, people will come. My fear is that it's going to be overcrowded and there will end up being a lot of clearing of the forest. It's a real concern – how populated it will be in years to come – and the ongoing impacts on the coastal environment as well the Reef.

There is a program we're working on with Traditional Owners called 'Strong Peoples — Strong Country'. It's a monitoring framework for important work to be done by the Traditional Owners and for them to tell their own stories of the work they are doing to maintain conservation and protection of the marine, coastal and reef biodiversity and its ecosystems.

Our Ancestors could tell what was happening in the ocean: particular plants that told them that a certain species of fish was around, for example. It was the Traditional Owners who could determine whether the particular fish were still plentiful or whether the numbers have gone down. They could monitor these movements and tell other Aboriginal people about this change.

It's important to ensure that Traditional Owners have the right information before any activity is started so that they can make an informed decision and provide their consent, and then actually empower themselves by doing some of the monitoring work themselves.

This is the sort of information that we're looking for now through monitoring and reporting through the 'Strong Peoples — Strong Country' tool. Hopefully through this project we'll find indicators that will help us to feed that information up from the community level, into the overall Great Barrier Reef Marine Park Authority monitoring system.

As a Traditional Owner, we have governments coming to consult with us, and they write up their reports. Those reports go to managing and monitoring agencies and sometimes it's not interpreted exactly as we say, and that's a problem. So, for the past couple of years there has been encouragement that once we get the information, we the Traditional Owners can control the information and write our own Traditional Owner reports to governments and other agencies so that they can work from our reports.

Chrissy Grant is an Aboriginal (Eastern Kuku Yalanji) and Torres Strait Islander (Mualgal from Moa Island) Elder. She has worked on GBRMPA projects, including the Strong Peoples — Strong Country Framework, and is a Member of the Tourism Reef Advisory Committee and the Reef 2050 Advisory Committee for the Great Barrier Reef.

Caring for colour: Multispecies aesthetics at the Great Barrier Reef

Killian Quigley
killian.quigley@acu.edu.au

Abstract

The Great Barrier Reef has been bleaching yet again. If the Anthropocene had a colour table, bleached coral would hold an especially recognizable place within it. By some lights, chromatic behaviour — and chromatic disaster — are best apprehended as secondary qualities, as spectacles that offer to point the discerning observer beyond the tokens of human sense and toward an object's (or ecosystem's) essential properties. This article asks whether it is possible, and ethically viable, to recognise corallian colour practice as having meaning in and of itself. I argue that we should recognise coral colourism as the irreducibly relational comportment of species, sunlight, salt water, sediment and so on. Contrary to some influential views, the Reef's performances are not simply constructed by the fantasies of human spectators, but by stimulating human sensoria, they do hail us as participants in the chromatic field. Reckoning the loss of hue as a discrete catastrophe might therefore generate tools for articulating value in a manner that is not strictly constructivist, naively scientific or reactionarily idealistic. Caring for the Reef may be, not first of all but not least of all, a caring for colour — a caring against chromatic disappearance and a caring towards chromatic repair.

> Deep where
> imperial volutes and hatchetfish live, colors humans have
>
> not yet named glow in caves made from black coral and clamshell.[1]

The enlargement of pallor

The Great Barrier Reef has been bleaching yet again. Along an unprecedented span of its 2,000 kilometre extent, aerial surveys conducted in early 2020 have disclosed the 'colorful' supplanted, in great swathes, by the 'white'.[2] For scientists working with the Reef, the enlargement of pallor furnishes empirical evidence that holobiontic relations — among polyps, algae, bacteria and others — are coming undone. This monumental unravelling serves, metonymically, as signal and foreshadowing of even deeper and wider cataclysms, from evolutionarily retrogressive ocean futures to planetary climate emergency. In the words of the anthropologist Cameron Allan McKean, 'Corals have become central figures of ecological collapse that reveal the ways anthropogenic climate change will unfold, unevenly, across the ocean.'[3] Lives, ecologies and aesthetics are in a tangle here, mutually assailed by the compounded anthropogenic violences at work on the Coral Sea. If the Anthropocene had a colour table, bleached coral would hold an especially recognizable — and symbolically resonant — place within it.

On the Reef and elsewhere, among corals, sponges, giant clams, human beings and other lives besides, the expression of colour is a dynamic multispecies performance. That performance's hues — and their absences — are manifestly interpretable as ciphers for ecological (ill-)health, and for physical and temporal proximity to stressors and stress events. 'The ghostly white spectacles of bleaching', as legal scholar Irus Braverman has it, 'are the mesmerizing face of the encroaching mass death in tropical corals, their human-induced but self-orchestrated requiem'.[4] To regard chromatic diminishment as the outward appearance of some more primarily existential phenomenon is to apprehend it as a kind of secondary quality, a representation that offers to point the discerning observer beyond the tokens of human sense and towards an object's (or ecosystem's) essential properties. Colour, on these terms, always stands for something else, and its vanishing is always the sign of some other, more fundamental privation.

Is it possible — and ethically viable — to tinker with this signifying chain and recognise multispecies colour practice as having meaning in and of itself? To ask this question is to force a useful reckoning with those currents in Western criticism that have subordinated sensory knowledge, and so left the field of so-called environmental aesthetics lacking in conceptual power. To answer this question affirmatively is to recognise coral colourism as the irreducibly relational comportment of species, sunlight, salt water, sediment and on. Contrary to some influential views, the Reef's performances are not simply constructed by the fantasies of human spectators who gaze (down) at them. But by stimulating human sensoria — and by generating forms of intersubjective pleasure — they do hail us as participants in the chromatic field. Reckoning the loss of hue as a discrete catastrophe might therefore generate fragile but hopeful tools for articulating 'ethico-aesthetic'[5] value in a manner that is not strictly constructivist, naively scientific or reactionarily idealistic. Caring for the Reef may be, not first of all but not least of all, a caring for colour — a caring against chromatic disappearance and a caring toward chromatic repair.

Colour and life

Thriving coral reefs are a synecdoche of marine biodiversity. It is often remarked that while they take up less than 1 per cent of planet Earth's surface area, they provide intermittent or permanent refuge for perhaps one-third of its oceanic creatures.[6] Such biotic munificence has been figured literally as aesthetic abundance by the artist David Liittschwager, who famously placed a one-cubic-foot 'biocube' on a reef crest in Moorea, in the South Pacific, and produced composite photographic portraits of the 'flourishing life' it framed.[7] So proceeding, Liittschwager and his pictures contribute to a longstanding tradition in Western environmental thought of conjoining the ideals of diversity and beauty so emphatically as to imply that they are mutually constitutive, if not actually synonymous. That the proponents of this tradition rarely meditate upon, let alone interrogate, this juncture may only testify to its presence and power.

Scleractinian, or reef-building, corals are cnidarian animals that have evolved to live symbiotically with miniscule algae called zooxanthellae, or zoox. Zoox take up residence by the millions inside coral polyps, where the algae photosynthesize nutrients from sunlight. Without this source of nourishment, scleractinians would lack the metabolic energy they need to construct reef systems. Moreover, and through a mechanism that remains incompletely understood by coral scientists, reefy colours are generated, intensified and diversified through interactions between corals and zooxanthellae, as well as with the 'vast array of microbial symbionts' that collectively render coral an entire community, or 'holobiont', properly speaking.[8] These relations, writes the marine biologist and author Helen Scales, 'give corals their bright colors and make life possible in parts of the ocean where nutrients are in short supply'.[9] Quietly but potently, the conjunction 'and' in Scales' phrase evokes a sense of ontological as well as temporal correspondence: the giving of colour and the making possible of life would seem inextricably linked here, in a fashion that does not obviously declare the primacy of one or the other operation.

'The colors are leaving.'[10] So writes McKean in a report from Japan's Yaeyama Island group, where reefs have been in marked decline ever since the first international bleaching event in 1998. When scleractinians are exposed to extreme marine heat events, coral polyps frequently eject their symbiotic zooxanthellae. This perversely ironic stress reaction, whereby an organism deprives itself of the relationship it requires to flourish, results in a startling 'loss of colour'.[11] This phenomenon, which McKean calls 'visually shocking',[12] has in recent years become a primary motif of environmental crisis, and specifically of anthropogenic warming. Widely-viewed documentary films, such as Jeff Orlowski's Emmy Award-winning *Chasing Coral* (2017), present vivid, deeply unsettling images of discolouration as ruination.[13] To make matters worse, it has been reported recently that the more 'structurally complex' a reef-building coral species is, the more devastating may be the impacts it suffers from bleaching.[14] Overheated, relationally subverted waters appear, on these terms, to be becoming not only less ecologically viable and less chromatically splendid but simpler, as though climate change were simultaneously uglifying the oceans and pushing them backwards in evolutionary time.[15] 'Anthropocene seas,' writes the environmental humanities theorist Stacy Alaimo, 'will be paradoxical, anachronistic zones of terribly compressed temporality where, it is feared, the future will move backwards'.[16]

'A thing is right,' wrote the conservation scientist Aldo Leopold, 'when it tends to preserve the integrity, stability, and beauty of the biotic community. It is wrong when it tends otherwise.'[17] With bleaching in view, the coincidence — or, rather, *entanglement* — of ecological and aesthetic value would seem superlatively pertinent to coralline ecosystems, and to the Great Barrier Reef. The marine scientist Ove Hoegh-Guldberg suggests as much when he articulates coral reef decline as a joint threat to 'biochemical heritage', to ecological productivity, and to the 'beauty of nature': a threatened species, he explains, is akin to 'an impressionist artist's work'.[18] The simile hinges on the idea that, like the destruction of an artistic masterpiece, the extirpation of a form of subaquatic life represents an absolute negation. A unique and unreproducible expression had existed in the world but no longer does, and this means the biome has suffered a loss that can never be recovered. On this view, the ecosystemic value of a species would appear to go beyond its contribution to biodiversity, quantitatively reckoned, toward something more deeply and more elusively inherent — something that cannot be abstracted from the species it characterizes, and that it is tempting to interpret as a kind of genius.

Some problems in coralline aesthetics

However compelling Hoegh-Guldberg's similitude may seem, and however appealing it may appear to liken corallian to human artfulness, comparisons like this one have a generally fraught track record across an array of discourses and disciplines, from environmental studies to international heritage law. It is simultaneously true, argues the philosopher Emily Brady, that 'aesthetic value' has fundamentally informed certain 'ecological values' — notably 'variety, diversity and harmony' — *and* is commonly dismissed as 'weak', 'trivial' and 'subjective' when invoked explicitly in the context of conservation policy debates.[19] If this is right, then aesthetics relates to ecology through a dynamic we might figure in terms of a form of repression: 'harmony, symmetry and integrity' are powerful concepts for thinking about ecosystems, but only if their constitutive 'aesthetic meaning' and the various vulnerabilities that meaning and its histories may expose are kept from view.[20] Such vulnerabilities are not difficult to imagine. To borrow an example close at hand, we might say that summoning a tradition as particular as Impressionist painting to elucidate the stakes of marine extinction demonstrates, in a flamboyant if not exceptional manner, the contingent formation of certain species' 'aesthetic' and 'symbolic significance'.[21] Ethical as well as intellectual questions spring up everywhere in this vicinity: what, after all, is the projection of culturally specific 'notions' of beauty if not yet another instance of eco-'cultural imperialism'?[22]

A diagnosis of just this kind has been issued recently in a critique of Western conservation modalities as conveyed among and upon the islands and waters of Melanesia. A tendency toward the 'aesthetic fetishization' of certain ecosystems — notably 'coral reefs' — has been interpreted as not only culturally blinkered but practically mistaken: the protection of marine beauty and biodiversity and the security of Melanesians' alimentary and economic wellbeing may be unrelated, if not actually opposed.[23] A growing chorus of voices have also begun making a related case *vis-à-vis* the Coral Sea. Along the Great Barrier Reef, writes the heritage scholar Celmara Pocock, it would be an error to mistake reefy aesthetics for an 'intrinsic characteristic' when such value really consists of a tissue of 'colonial myths', scientific and technological discourses, tourist imaginaries and Western conventions in art-making and art appreciation, especially as they have pertained to sublimity and landscape theory.[24] Along similar lines, the art historian Ann Elias has recently observed the cultivation, around the middle of the twentieth century, of a particular image of the Reef's aesthetic value. Elias situates this image, which stressed the Reef's astonishing 'variety of color and form', directly within the promotion and protection of national interest, as governments in Australia and elsewhere sought to establish their subsea territories as most dazzling, and so most appealing to international travellers.[25]

By constructing their subject and its 'visual allure' as a 'glamorous commodity', Elias argues, the architects of what may be the dominant mode of looking at the Reef drew upon and reproduced older and wider conventions in aestheticizing tropical nature and tropical place.[26] This analysis establishes the Reef as one node in a broader network of colonial and quasi-colonial entanglements, where aesthetic, economic and scientific meanings were co-constituted by imperial actors in sites such as the Caribbean.[27] It resonates with such formative strains in recent ecocritical and environmental-humanist theory as have cautioned against a 'logic of Romantic consumerism'[28] that may continue to structure natural and wilderness ideals — and may thwart needed work towards ecological repair and environmental justice. What is emerging to view here is a sense of the aesthetic as not only problematically subjective, and therefore 'non-epistemic', but deeply and pervasively compromised by its cultural and historical genetics.[29] This is a vital reckoning, and a damning one too: if these forms of value are already mouldering, as ill-defined 'cultural services'[30] and the like, at the shadowy margins of conservation discourse, perhaps the best thing to do is to let them crumple into dust like so many regrettable, forgettable ruins.

To imagine as much is to imply that these energies are fully and firmly within our grasp — that just as certain among us, and among our cultural antecedents, have cobbled the aesthetic together, so might we now reckon its joints, take it apart, and perhaps leave its fragments well aside. The question I am attempting to pose, along with numerous others, is whether this account of things is sufficient or right. I hypothesise that the answer to the question is 'no', and that we know this from the planetary impoverishment coral bleaching effects, an impoverishment that can be partly contextualized but never fully accounted for by means of an hermeneutics of suspicion.[31] The colours are leaving, and they are doing so well beyond the jurisdictions, park authorities and even cultural imaginaries of any particular national or imperial actor, however extensively and disproportionately those actors may have contributed to the socioecological disasters of our pasts, presents and futures. The Anthropocene, writes the novelist and theorist Amitav Ghosh, 'is precisely a world of insistent, inescapable continuities, animated by forces that are nothing if not inconceivably vast'.[32] One of those continuities is the expulsion of coral colour, and one of our concerns is what we will manage to say about, and feel for, that colour's extinguishment, as well as the pallid nothing it leaves behind.

Sensation, beauty, and scientific cognitivism, submerged

In his essay on the aesthetics and politics of the bleaching Great Barrier, the literary historian Jonathan Lamb coordinates philosophical suspicion of perceptual information with contemporary negligence for colour-care. An attitude of 'resistance to sensation' in Western epistemology, Lamb writes, subtends a perversely solipsistic view wherein abstract ideas acquire more substance than a world that only suggests them — and wherein the apparent loss of reefy colour is just that: mere appearance.[33] I argue that this habit of thought remains unresolved amidst conventional theories of environmental aesthetics in the West. It animates, on the one hand, a tenacious sense of aesthetics as concerned above all with the beautiful in nature, and the beautiful as either overdetermined (and therefore compromised) by perceptual idiosyncrasy and cultural preference or as transcendentally apolitical. Obliquely, but no less significantly, it also informs the scientific-cognitivist view that an appropriately rigorous environmental aesthetics must derive from an observer's pre-existing familiarity with pertinent conceptual — that is, primary — knowledge.

Justifying the Reef's inclusion in the rolls of World Heritage, the United Nations Educational, Scientific and Cultural Organization cites its 'superlative natural beauty above and below water,' its provision of 'some of the most spectacular scenery on earth,' its 'unparalleled aerial panorama of seascapes comprising diverse shapes and sizes,' the 'myriad of brilliant colours, shapes and sizes' exhibited by its denizens, and so on.[34] By characterizing its subject's aesthetic value in these terms, UNESCO trades in an overtly European vocabulary of the beautiful, the theatrical, the picturesque, and the variegated. As Pocock astutely observes, the assumptions and concepts undergirding the Reef's World Heritage designation are thoroughly intertwined with ocularcentrism, as well as with 'Western notions of natural and cultural aesthetics' more broadly.[35] Projected upon holobiontic lives, these notions also reveal themselves, upon scrutiny, as bizarrely 'terracentric', to adapt the historian Marcus Rediker's term.[36] Among marine and submarine environments where the logics of scenery and panoramas operate weirdly, if they can be said to operate at all, World Heritage declines to acknowledge, let alone extensively consider, the distinctive aesthetic affordances and disruptions of subsea space.[37] The undersea is a 'world', in the architectural theorist William Firebrace's words, 'with its own aesthetics'.[38] So far as human beings are concerned, that world is defined by nothing so much as its 'presenting entirely different conditions for perception, sensation, and life than terrestrial environments', as the media scholar Melody Jue wrote.[39] Representing the Reef as a series of spectacular pictures, then, not only hitches what aesthetic value it might be said to hold to a specific and fraught lineage, but actually misses what are arguably the most foundational aspects of its aquatic and subaquatic character.

The case for beauty has, ironically, been undermined further by critics whose efforts to defend it may be characterized at best by conceptual imprecision and at worst by a kind of reactionary quietism. The aesthetic, claims the introduction to a recent volume on *Ecocritical Aesthetics*, is a 'sober-sounding, opaque' and basically 'sterile' category advanced in the latter decades of the twentieth century to appease (or evade) the intellectual and political demands of those 'postmodern' theorists for whom the beautiful had been revealed as fundamentally, ideologically fraught.[40] What goes unstated in an account like this one is the historical process whereby, as the visual culture theorist Nicholas Mirzoeff has it, a capacious sense for 'aesthesis', or the 'full range of the body's sensorium', became constricted over the course of the eighteenth century to a concern for the beautiful in art.[41] To figure the replacement of beauty by aesthetics as the work of timorous critical accommodationists is to erase the history Mirzoeff tells, as well as to conflate two historically distinctive terms. And later on in *Ecocritical Aesthetics*, the political stakes of this supposed substitution are rendered more, and more troublingly, explicit by a revealing question. 'Can we "resurrect" beauty', asks one ecocritic, 'in the face of two generations of the steady operationalization of literature and culture as instruments of the identity politics of marginal groups?'[42] By proffering so narrow and tendentious an impression of what an environmental aesthetic is and has been, and by configuring it as inherently threatened by the epistemological and institutional claims of (caricatured) 'marginal groups', these notions construct a domain of thought and practice that few serious students of ecocriticism will recognize, and fewer still will find hospitable.

What World Heritage and these (which is by no means to say all) ecocritical aestheticians have in common is an uncritical, and sometimes explicitly ideological, commitment to the concept of beauty, as well as what Lamb characterizes as a resistance to sensation. The latter tendency also typifies some versions of an environmental aesthetics that Brady identifies as 'scientific cognitivism'. By scientific cognitivist lights, 'natural aesthetics' must be informed by the 'natural sciences', so sensory impressions are at best some means toward, and never autonomous engines of, meaningful perceptual knowledge.[43] As the philosopher Jennifer Welchman observes, the scientific-cognitivist position is calculated as an alternative to a 'formalist' approach to the aesthetic appreciation of nature, in order especially to obviate that approach's vulnerability to accusations of being 'purely subjective'. Toward a kind of convergence between aesthetic and ecological epistemologies, the scientific-cognitivists propose, in effect, that the more sophisticated one's scientific comprehension of a life, a place, or a process, the more apt one's apprehension of its beauty.[44] From among these anxieties, vexations and real problems, the path towards a theory of coral colourism appears poorly envisioned, let alone trod, by certain extant paradigms. Fortunately, certain others await enlistment to the cause at hand.

Prismatic ecologies of the Great Barrier Reef

Where to turn for an environmental aesthetics willing to credit that the stirrings of colours and other stimuli might be something other than inherently epistemically insignificant? In the next several paragraphs, I hope to take this discussion in a more generative direction by coordinating our reefy discussion with the related works of a number of recent theorists of colour, some of whose works have been assembled in a rich collection of essays titled *Prismatic Ecology*. What the prismatic ecologists and some related thinkers share is a commitment to understanding colour as a kind of emergent property of bodies in relation: 'when light, objects, eyes, and perceiving beings engage,' writes the environmental humanities scholar Heather I. Sullivan in a study of Johann Wolfgang Goethe, 'then colors emerge'.[45] By holding colour in view — colour considered not as a static picture but as a shifting, elusive process — as one among a host of other primary terms, a prismatic aesthetics locates epistemological and affective meaning with and in chromatic phenomena.

'Color,' argues the literary scholar Jeffrey Jerome Cohen, 'is not some intangible quality' of an encounter 'but a material impress, an agency and partner, a thing made of other things through which worlds arrive'.[46] Cohen's invocation of materiality draws attention to the ways colourism exists in the world and configures hosts of relations, or what Sullivan calls 'ecolog[ies] of color and light', incorporating 'human beings, bodies, cultures, practices, and the vast array of more-than-human biotic and abiotic factors'.[47] More than the mere face of primary phenomena, colour appears here as endowed with actualities, and with world-making potentialities, unto itself. In this regard, Cohen's, Sullivan's and others' writings resonate with the anthropologist Eduardo Kohn's account of the rainforest around Ávila in Ecuador. The forest's cacophonous, more-than-human communities achieve, Kohn writes, a 'relatively more nuanced and exhaustive overall representation of the surrounding environment when compared to the way life represents elsewhere on the planet'.[48] The normative implications of Kohn's account are complex and exciting. What might it mean to celebrate, and to defend, a biodiverse rainforest as the realm of exemplary more-than-human representation? To regard an abundant reef as the domain of exquisite multispecies colour?

What is radically useful about a prismatic ecology is that it thinks through colour to simultaneously reckon the ineluctable role of the perceiver in the chromatic moment *and* grapple with the fact that what constitutes colour always exceeds human apprehension. Material culture scholar Julian Yates contends that colour 'might best be modeled as a multispecies sensory process or network that generates biosemiotic-material effects that then take on a metaphorical life of their own as they are translated to different registers'.[49] This account offers a usefully multidimensional frame for comprehending chromatic presence, one that does not rely on the dichotomy of that which is either wholly 'dependent on our perception' or absolutely 'a property of [an] object itself'.[50] And multiplicity is precisely what is needed to comprehend a phenomenon like bleaching at the Great Barrier Reef, a situation that both intersects powerfully with a Western gaze and the histories that have constituted it *and* exceeds that gaze in undeniable ways – thus, I think, creating a productive (and by no means unadmirable) tension at the heart of Elias's book. On the one hand, her text links the fetishization of colour directly to ideologies of conquering the sea, of turning it into so much 'resource' to be extracted and exploited. On the other, it is movingly concerned with the fate of (among countless other creatures) the damselfish, which as Elias explains requires 'a healthy coral reef; a profusion and abundance of forms and colors;', a 'habitat where animal, vegetable, and mineral are difficult to distinguish and where even the most dramatically colored species can hide in plain sight and not stand out like specimens in an aquarium'.[51]

Human regard *for* colour and the fundamental preciousness *of* coral colourism are thinkable, then, as simultaneous, and in some regards entangled, but not identical aspects of the Reef's meaning. This matters, because it suggests an opening towards — and a responsibility for — valuing and cultivating reefy colour that do not oppose but *accompany* a reckoning with the ways marine aesthetics have been misrepresented, commoditized and so forth. 'Viewing our enmeshment and ongoing engagements' as integral to a broader 'ecology of color', contends Sullivan, 'might inspire an awareness of our material circumstances that we impact by being alive'.[52] Articulating the injury done through anthropogenic bleaching, and advocating on behalf of corallian splendour, need not entail a fetishization of narrowly contingent cultural value. Nor, for that matter, need it necessarily resort to a scientistic explication of the ecological realities that lie 'behind' chromatic appearance. A 'vast coral reef' is the ecological theorist Timothy Morton's metaphor for all the 'actually existing beings' that really subtend phenomenological experience but that are (as Lamb observed) too readily explained away as only apparitions. With Morton's figure — 'strange, withdrawn' and 'yet forever sending out signals' such as 'form, color, spatiality' — the potent if only semi-scrutable nature of reefy colourism demonstrates its capacity for moving through language as well as through registers of being, structuring and restructuring possible worlds in manners observable and not.[53]

Colour-care

At the Reef, among phenotypically promiscuous multispecies communities famous for frustrating the taxonomic eye, the protocols of classification yield to more immediate, and arguably less certain, modes of impression and description. A colony's colour practice may simply decline to contribute materials towards identification, or indeed towards story. The undersea, wrote the philosopher Alphonso Lingis, 'is all in surface effects'.[54] What Lingis described, and what coral colourism invites, is a lingering with the external qualities of things that is not *en route* to some more penetrative narration or illumination. At the Reef, after all, it is exactly at such superficies that lives are showing and colours are taking place.

Submarine realms work curious effects on human senses, and put distinctive pressures on the forms of human-involved relationality that characterize many terrestrial situations. To dive on the Reef is to recognize, not least of all, the contingent and coalitional nature of colour, as zooxanthellae algae and their symbionts stimulate my imperfect eyes. Even buttressed by my dive goggles' lenses, my vision down below is poor, particularly as it pertains to the chromatic spectrum. Thus one small instance of the paradoxes of immersed aesthetics: amidst hues of extraordinary diversity and brilliance, a submerged spectator experiences a radical diminishment in their ocular faculties. These colours simply do not exist *for* the sake of human observation — or as the projections of human fantasists.

Here, nonetheless, as the Australian artist Janet Laurence shows, is a potential site for primary care. Laurence's installation *Deep Breathing: Resuscitation for the Reef* (2016) figures among a panoply of other interventions a course of colour transfusions for coralline patients. Refusing the sort of logic that comprehends hue as just the ornamental emblem of some more primary reality, *Deep Breathing* suggests that concern for colour might offer an actual, if necessarily imperfectible, conduit to care.[55] The Reef's chromatic collaborations are foundationally more-than-human. At the same time, its performances make themselves known to human sense, and so available to ethical attention. Better mourning their losses, and better supporting their flourishing, may contribute to a species of colourism that is not the sign, but the stuff, of life.

Notes

1 Aimee Nezhukumatathil, 'Invitation', in *Oceanic* (Port Townsend, WA: Copper Canyon Press, 2018), p. 29.

2 Damien Cave, 'Great Barrier Reef is bleaching again. It's getting more widespread', *The New York Times*, 6 April 2020, https://www.nytimes.com/2020/04/06/world/australia/great-barrier-reefs-bleaching-dying.html.

3 Cameron Allan McKean, 'The horror of a rubble Reef', in Cameron Muir, Kirsten Wehner and Jenny Newell (eds), *Living with the Anthropocene: Love, loss and hope in the face of environmental crisis* (Sydney: NewSouth Publishing, 2020), p. 208.

4 Irus Braverman, *Coral whisperers: Scientists on the brink* (Oakland, CA: University of California Press, 2018), p. 79.

5 Félix Guattari, *The three ecologies*, trans. Ian Pindar and Paul Sutton (New Brunswick, NJ: Athlone Press, 2000), pp. 36–7.

6 Bärbel G. Bischof, 'Geographies of coral reef conservation: Global trends and environmental constructions,' in Jon Anderson and Kimberley Peters (eds), *Water worlds: Human geographies of the ocean* (Farnham: Ashgate, 2014), p. 53.

7 'Biocube in Moorea, French Polynesia', Smithsonian National Museum of National History, Washington, DC, https://naturalhistory.si.edu/education/teaching-resources/life-science/biocubes-exploring-biodiversity/biocubes-action/moorea.

8 'Biocube in Moorea', p. 7.

9 Helen Scales, 'From polyp to rampart: The science of reef building and how art can inspire a sustainable future', in Jason deCaires Taylor, *The underwater museum: The submerged sculptures of Jason deCaires Taylor* (San Francisco: Chronicle Books, 2014), p. 19.

10 Cameron Allan McKean, 'The color of climate change', *The Japan Times*, 4 March 2018, 10.

11 Terry P. Hughes et al., 'Ecological memory modifies the cumulative impact of recurrent climate extremes', *Nature Climate Change* 9 (2019), 44.

12 McKean, 'The color of climate change', 10.

13 Jeff Orlowski (dir.), *Chasing Coral*, Netflix, 2017.

14 James Bradley, 'The end of the oceans', *The Monthly*, August 2018, https://www.themonthly.com.au/issue/2018/august/1533045600/james-bradley/end-oceans.

15 For a discussion of future oceans as biologically simpler, see Jan Zalasiewicz and Mark Williams, *Ocean Worlds: The story of seas on Earth and other planets* (Oxford: Oxford University Press, 2014), p. 191.

16 Stacy Alaimo, 'The Anthropocene at sea: Temporality, paradox, compression', in Ursula K. Heise, Jon Christensen and Michelle Niemann (eds), *The Routledge companion to the environmental humanities* (London: Routledge, 2017), p. 158.

17 Aldo Leopold, *A Sand County almanac: With essays on conservation from Round River* (New York: Ballantine Books, 1982), p. 262.

18 Braverman, *Coral whisperers*, pp. 58–9.

19 Emily Brady, 'Aesthetics in practice: Valuing the natural world', *Environmental Values* 15 (2006), 278–9.

20 Brady, 'Aesthetics in practice', 283.

21 Dirk Lanzerath, 'Biodiversity as an ethical concept: An introduction', in Dirk Lanzerath and Minou Friele (eds), *Concepts and values in biodiversity* (London: Routledge, 2014), p. 11.

22 William Cronon, 'The trouble with wilderness; or, getting back to the wrong nature', in William Cronon (ed.), *Uncommon ground: Rethinking the human place in Nature* (New York: W.W. Norton, 1996), p. 82.

23 Simon Foale, Michelle Dyer and Jeff Kinch, 'The value of tropical biodiversity in rural Melanesia', *Valuation Studies* 4 (2016), 13, 24.

24 Celmara Pocock, *Visitor encounters with the Great Barrier Reef: Aesthetics, heritage and the senses* (London: Routledge, 2020), pp. 1, 6.

25 Ann Elias, *Coral empire: Underwater oceans, colonial tropics, visual modernity* (Durham, NC: Duke University Press, 2019), p. 203.

26 Elias, *Coral empire*, 207, 210.

27 Megan Raby, *American tropics: The Caribbean roots of biodiversity science* (Chapel Hill, NC: University of North Carolina Press, 2017), pp. 67–8.

28 Timothy Morton, *Ecology without nature: Rethinking environmental aesthetics* (Cambridge, MA: Harvard University Press, 2007), p. 113.

29 David M. Frank, '"Biodiversity" and biological diversities: Consequences of pluralism between biology and policy', in Justin Garson, Anya Plutynski and Sahotra Sarkar (eds), *The Routledge handbook of philosophy of biodiversity* (London: Routledge, 2016), p. 100.

30 Keith Hiscock, *Marine biodiversity conservation: A practical approach* (London: Routledge, 2014), p. 10.

31 I borrow this phrase from Paul Ricoeur via Rita Felski, 'Suspicious minds', *Poetics Today* 32 (2011), 216.

32 Amitav Ghosh, *The great derangement: Climate change and the unthinkable* (Chicago: University of Chicago Press, 2016), p. 62.

33 Jonathan Lamb, 'Understanding the loss of colour', in Margaret Cohen and Killian Quigley (eds), *The aesthetics of the undersea* (London: Routledge, 2019), p. 56.

34 'Great Barrier Reef', UNESCO World Heritage Centre, http://whc.unesco.org/en/list/154.

35 Celmara Pocock, 'Sense matters: Aesthetic values of the Great Barrier Reef', *International Journal of Heritage Studies* 8 (2002), 380.

36 Marcus Rediker, 'Hydrarchy and terracentrism', in Alex Farquharson and Martin Clark (eds), *Aquatopia: The imaginary of the ocean deep* (Nottingham: Nottingham Contemporary and London: Tate Publishing), p. 115.

37 For an extended discussion of this distinctiveness, see Margaret Cohen and Killian Quigley, 'Submarine aesthetics', introduction to Cohen and Quigley (eds), *The Aesthetics of the Undersea*, pp. 1–13.

38 William Firebrace, *Memo for Nemo* (London: AA Publications, 2016), p. 65.

39 Melody Jue, *Wild blue media: Thinking through seawater* (Durham, NC: Duke University Press, 2020), p. 10.

40 Peter Quigley, introduction to Peter Quigley and Scott Slovic (eds), *Ecocritical aesthetics: Language, beauty, and the environment* (Bloomington, IN: Indiana University Press, 2018), pp. 3, 15.

41 Nicholas Mirzoeff, 'Visualizing the Anthropocene', *Public Culture* 26 (2014), 219. For further discussion of debates respecting the eighteenth-century origins of an 'aesthetic theory of art', see also Paul Guyer, 'History of modern aesthetics', in Jerrold Levinson (ed.), *The Oxford handbook of aesthetics* (Oxford: Oxford University Press, 2005), pp. 29–30.

42 Mark Luccarelli, 'Renaissance aesthetics, picturesque beauty, the natural landscape: An essay examining the rise and fall of the impulse toward beauty', in Quigley and Slovic (eds.), *Ecocritical aesthetics*, p. 80.

43 Emily Brady, 'Aesthetic value, nature, and environment', in Stephen M. Gardiner and Allen Thompson (eds), *The Oxford handbook of environmental ethics* (Oxford: Oxford University Press, 2016), p. 188.

44 Jennifer Welchman, 'Aesthetics of nature, constitutive goods, and environmental conservation: A defense of moderate formalist aesthetics', *The Journal of Aesthetics and Art Criticism* 76 (2018), 419–21.

45 Heather I. Sullivan, 'The ecology of colors: Goethe's materialist optics and ecological posthumanism', in Serenella Iovino and Serpil Oppermann (eds), *Material ecocriticism* (Bloomington, IN: Indiana University Press, 2014), pp. 80–1.

46 Jeffrey Jerome Cohen, 'Ecology's rainbow', introduction to *Prismatic ecology: Ecotheory beyond green* (Minneapolis, MN: University of Minnesota Press, 2013), p. xvi.

47 Sullivan, 'The ecology of colors', 89.

48 Eduardo Kohn, *How forests think: Toward an anthropology beyond the human* (Berkeley, CA: University of California Press, 2013), p. 81.

49 Julian Yates, 'Orange', in Cohen (ed.), *Prismatic ecology*, p. 85.

50 Vittoria di Palma, 'A natural history of ornament', in Gülru Necipoğlu and Alina Payne (eds), *Histories of ornament: From global to local* (Princeton, NJ: Princeton University Press, 2016), p. 26.

51 Elias, *Coral empire*, pp. 212–13.

52 Sullivan, 'The ecology of colors', 93.

53 Timothy Morton, 'X-Ray,' in Jeffrey Jerome Cohen (ed), *Prismatic Ecology: Ecotheory Beyond Green* (Minneapolis and London: University of Minnesota Press, 2013), p. 312.

54 Alphonso Lingis, 'The rapture of the deep', in *Excesses: Eros and culture* (Albany, NY: State University of New York Press, 1983), p. 7.

55 Janet Laurence and Prudence Gibson, 'The ocean hospital – a walk around the ward', in Cohen and Quigley (eds.), *The aesthetics of the undersea*, pp. 198–9.

Killian Quigley is a research fellow at the Institute for Humanities and Social Sciences, Australian Catholic University and honorary postdoctoral fellow at the Sydney Environment Institute, University of Sydney. He is the co-editor, with Margaret Cohen, of *The Aesthetics of the Undersea* and author of the forthcoming *Reading Underwater Wreckage: An Encrusting Ocean*. His research is available, now or imminently, from *Environmental Humanities*, *Green Letters*, *A Cultural History of the Sea in the Age of Enlightenment*, *Maritime Animals: Ships, Species, Stories* and elsewhere. He is an associate member of the Oceanic Humanities for the Global South research network.

Coralations: Back to the breath

Irus Braverman
irusb@buffalo.edu

> You and me
> Knew life itself is
> Breathing,
> (Out, in, out, in, out ...)
> Breathing
> – Kate Bush, 'Breathing, on *Never for Ever* (1980)

Figure 1.
Immersing and breathing in the Great Barrier Reef. Photo by author, 2015.

Corals are good to breathe with. Living painfully far from the ocean during the long COVID-19 lockdowns, I have been relegated to daydreaming about being immersed in salty waters again. As a way to feel closer, I succumbed to shuffling through the amateur videos and photos I shot in various diving locations around the world: the Caribbean, the Red Sea, Hawai'i. Some way into the shuffling, I stumbled upon images from my two-day visit to the Great Barrier Reef, off Port Douglas, in

2015. This visit occurred just before the massive third global bleaching event, which decimated corals on a massive scale. I was lucky to have visited the Reef before it lost so much of its vibrancy.

I would like to dedicate this contribution to a concept I came up with when contemplating the vibrant interrelations among various coral parts as well as between corals and other living beings: 'coralations'.

The symbiotic algae–microbes–animal relationship at the core of the corals' precarious existence reveals that, more than a single unified entity, corals are 'coralations' — bundles of constantly changing associations that shape and reshape their ways of being in the world, and therefore the world itself. Beyond the symbiotic underpinnings of their microscalar existence, coralations also occur at the level of the coral colony and ecosystem, and then at the intersections of culture, science and law.

Exploring coralations involves immersing ourselves and breathing with the corals. It calls us to tune into our sensorial dimensions, thus allowing ourselves to experience and be with them, not just analyse them from afar. This concept foregrounds the deep entanglements of human–non-human relationships, while inviting us to breathe life and hope to translate such coralations into action.

Alongside breathing with corals, there is also the importance of 'thinking with' corals. Although they are more apparent in liquid environments, where the dynamic interconnections and flows are easier to see and sense, coralations can encompass a range of legal and scientific spaces.

When such thinking happens, corals confuse and destabilise our categories: they are a cross between animal, plant, rock, microbe and ecosystem; we sentimentalize corals because of their beauty, despite the fact that they don't have a face, or a clear sex, so we can't easily anthropomorphise them; and while they live in the ocean, which constitutes the majority of the Earth but which we know so little about, they also constitute some of our terrestrial mountains and buildings.

Furthermore, reef building corals are animals yet they photosynthesize; they make massive stony structures that can be seen from space but they are tiny and, some claim, fragile creatures; they are sessile yet travel long distances in their larvae stage; and each has a mysterious symbiotic relationship with a particular strain of algae, which under certain conditions disembarks from the coral cells, 'leaving' them white and depleted (or 'bleached'). 'Individual' coral polyps in a colony — scientists are not in agreement about what individual corals are — may differ in morphology and genetics, and some may be fusions of two or more genotypes. For the most part, however, polyps who belong to one colony have the same genetic composition, what scientists refer to as 'ramets'. Coral colonies are interconnected by living tissue.

Because of their hybrid materialities, corals have long inspired Indigenous cultures, poetry and art. The Kumulipo ('Beginning-in-deep-darkness') is the sacred creation chant of a family of Hawai'ian ruling chiefs. Composed and transmitted entirely in the oral tradition, its 2,000 lines provide an extended genealogy proving the family's divine origin and tracing its history from the beginning of the world. The Kumulipo opens with the coral as the first organism in the universe. According to the Kumulipo, corals are the beginning of life, the most ancient ancestors of all living things. Unfortunately, in recent times corals have increasingly been embodying and signalling the possible end of life. This is yet another type of coralation.

Figure 2.
Corals and humans from a fish's perspective. Photo by author, Angel Reef, Great Barrier Reef, 2015.

And back to thinking with corals. Corals have fascinated great intellectuals such as Karl Marx, who mentions them as prime examples of the relationship between the individual and the community. Charles Darwin's first monograph in 1842 was entitled *The Structure and Distribution of Coral Reefs, Being the first part of the geology of the voyage of the Beagle, under the command of Capt. Fitzroy, R.N. during the years 1832 to 1836*. More recently, the coral made a prominent appearance in evolutionary biologist and historian of science Stephen J. Gould's last book, *The Structure of Evolutionary Theory*. There, Gould used the picture of a coral to represent the basic tenets of Darwinian theory.

Ironically, many scientists who have studied corals have come to challenge the traditional Darwinian principles of evolution, highlighting the centrality of the symbiotic relationship and the importance of understanding the coral as a 'holobiont' (again, a composite of coral animal, algae and a diverse set of microbes). These coral-spawned realisations have brought about a substantial paradigm shift in the field of biology, which until recently has been dominated by neo-Darwinian theories of origin and natural selection. It is thus not very surprising that scientists who study corals also tend to promote a 'rhizomatic' outlook of the world. The 'rhizome' is a concept developed by French philosophers Gilles Deleuze and Felix Guattari in the 1970s to highlight ways of thinking that are multiple and nonhierarchical, as opposed to 'arborescent' (tree-like and hierarchic) knowledge that works with dualistic categories and binary choices. As Darwin himself acknowledged in his notebook, 'The tree of life should perhaps be called the coral of life', implying that his own view was much less Darwinian than it was later interpreted to be.

In the Anthropocene, the project of managing coral can also be understood as a coralation, this time of law and science. My book *Coral Whisperers: Scientists on the Brink* uncovers myriad mundane ways in which coral life and law are entangled — law breathes life into corals, if you will, by protecting their continued existence, and the coral breathes life into law by providing it with the materiality required for governance. Indeed, corals are saturated with law, which protects, ignores and damages them, or any combination thereof; they are also saturated with science, which names, classifies and lists them, thereby making them legible, or illegible, to the law. The term 'coralations' encompasses the co-productive elements of, and dynamics between, law and science, alongside less mutualistic relations that include tension and even negation.

For now, though, all I can do is shuffle through photos, write and dream about coralations, always breathing – in, out, in.

Irus Braverman is Professor of Law and Adjunct Professor of Geography at the University at Buffalo, State University of New York. Her books include *Zooland: The Institution of Captivity* (2012), *Wild Life: The Institution of Nature* (2015) and *Coral Whisperers: Scientists on the Brink* (2018).

Aquariums and human–animal relations at the Great Barrier Reef

Ann Elias
ann.elias@sydney.edu.au

Abstract

In the early twentieth century, great delight in the unique tropical beauty of the Great Barrier Reef, coupled with an opportunistic spirit for commercial development, inspired the commission of eye-catching posters and advertisements by Australian tourist organisations. The aim of this article is to discuss a pictorial device that developed alongside the rise of modern tourist advertising images of Great Barrier Reef – a split-level viewpoint that approximates the effect of looking at the Reef through the glass sides of an aquarium. Building on my earlier research published in 2019 on wildlife photography and the construction of the Great Barrier Reef as a modern visual spectacle, and combining art history with environmental history, this article also turns to coloured advertising lithographs. It argues that split-level visualisations separate human from non-human and elevate the idea of human superiority. With the Great Barrier Reef facing unprecedented ecological pressures, the historical images at the centre of this article are instructive for understanding the deleterious effects of anthropogenic impact, as well as early twentieth-century attitudes towards human–non-human relations.

Introduction

In the early stages of researching a book on the significance of the Great Barrier Reef to modern visual culture, I came across a provocative comment by photographer and explorer Frank Hurley (1885–1962).[1] Having peered over the side of a boat in 1921 to look at a submerged coral reef in North Queensland, Hurley felt mesmerised by the tropical underwater scene that unfolded before his eyes and described the effect as 'much like looking into a glorified aquarium'.[2] I was struck by his comment's domestication of the wild flora and fauna of the Great Barrier Reef and by the anthropocentric nature of the analogy of the sea to a human invention that miniaturises the sea and, in domestic contexts, puts animals on display as decorative objects.

It became apparent after further research that Hurley's analogy of a coral reef to an aquarium was far from an isolated case. Celmara Pocock also notes an instance in 1935 when the Reef was described as 'a marvellous aquarium. Now look at the exhibits: there's a dogfish … a bêche-de-mer …'.[3] As my research into visual representations of the Great Barrier Reef progressed, three key themes began to emerge: exploitation of the Reef for tourism; nature as spectacle; and the visual objectification of marine animals. With this article, I return to these themes and to the metaphor of the Great Barrier Reef as an aquarium. But whereas my previous research focused on photography, this article concentrates on coloured lithographs produced for posters, mass production and tourist advertising. Having reached audiences worldwide they have been a significant influence on popular understandings of the Great Barrier Reef. Rather than focus on how the images sold Australia, which is how Michelle Hetherington has framed the visual propaganda of tourist posters, my intention is to relate them to societal attitudes to non-human animals.[4] In sum, I believe the images in this article reveal why the ethics of human–non-human relations is one of the pressing issues of our time. These images celebrate the remoteness and uniqueness of nature at the Reef, but also embody environmentally unfriendly attitudes to animals. Recently, political ecologist Jane Bennett challenged readers of *Vibrant Matter: A Political ecology of Things* to 'picture an ontological field without any unequivocal demarcations between human, animal, vegetable, or mineral'.[5] The images in this article, however, encourage readers to do the opposite of Bennett's suggestion by clearly demarcating the human realm from non-human realms. They are evidence of how the visual culture of the past can offer insights into attitudes that contemporary societies need to overcome if we are to slow the process of extinctions of animals at the Great Barrier Reef, including corals and fish.

At the centre of this discussion are four tourist advertisements (Figures 1–4), three of which are from the 1930s. I argue that they exhibit an extractive mentality and an anthropocentric mindset that can be linked to a failure of care for the Reef and its animals. As W.J.T. Mitchell argues, visual images — whether scientific illustrations, paintings from fine art, or colour lithographs used for tourism — are 'symbolic constructions ... that interpose an ideological veil between us and the real world'.[6] They mediate knowledge and are never self-evident, but rather are puzzles to analyse and deconstruct. How has the visual practice of advertising impacted the Great Barrier Reef? Through examples, I argue that modern advertisements separated 'nature' from 'culture' and privileged human life and human activity.

The discussion concentrates on a specific pictorial device that emerged in the 1930s in travel advertisements for the Reef. In this article, it is defined as a 'split-level viewpoint'. This viewpoint approximates the effect of looking at the Reef as if through the glass sides of an aquarium, and of seeing in one pictorial frame a view below water as well as one above. The split-level view simulates an imagined fish-eye viewpoint from below, yet it is never suggested that the viewing position is an immersed one; rather, it is suggested that the viewpoint is located outside the picture, in air, like the eye of a viewer on land observing an aquarium from behind glass.

I relate this form of visualisation to its grounding in a biopolitical hierarchy: the modern ontological divide between human and animal. Randy Malamud argues that animals are malleable to whatever image and identity people want to make of them.[7] The split-level perspective produces a heightened sensation that marine animals and people are alien to each other and exist in realms that are separate and distinct, rather than living in a multi-species planetary context. It highlights Judith Hamera's point in *Parlor Ponds* that aquarium viewpoints encourage the cultural

Figure 1
Percy Trompf, 'The Marine Wonders of the Great Barrier Coral Reef' (Queensland Government Tourist Bureau), 1933, 100 × 63.8 cm, Queensland State Archives.

Figure 2
The Great Barrier Reef, Queensland, the sunshine state of Australia, 1958–1965 (Queensland Government Tourist Bureau), colour lithograph, National Library of Australia.

viewpoint of a distanced position from the animals of the underwater.[8] It produces the idea that animals are symbolically far away from the viewer despite being very close.

The article begins with an account of the rise of the Great Barrier Reef as an iconic geography perceived to have significant potential for exploitation and tourism. It then turns to Figures 1 and 2 to discuss the spectacle of the optics of

Figure 3

Left: James Northfield, Australia: Great Barrier Coral Reef, c. 1932, Australian National Travel Association, colour lithographic poster 101.7 x 63.8 cm, National Library of Australia, PIC Poster Drawer 289.

Figure 4
Noel Pascoe Lambert, The anglers' paradise, Great Barrier Reef, 1936, Queensland Government Tourist Bureau, colour lithographic poster, National Library of Australia.

tourist images of the Reef and specifically the split-level viewpoint and its relation to aquarium modes of seeing.[9] A section follows that interrogates the symbolic value of the split-level viewpoint and proposes that fear of the animal Other undergirds a predilection for the distancing effect of an aquarium. The discussion then turns to Figures 3 and 4, arguing that a connection between the visual strategy of radically exposing the bodies of marine animals to the gaze of the viewer and a history of exploitation and diminished care for the marine animals of the Reef existed in an era that thought the riches of the seas were boundless.

Exploitation and the rise of tourism

By 1929, it was widely recognised that the Great Barrier Reef's chief future industry would be tourism, in particular a tourist industry based in scientific curiosity about corals, the visual pleasure of coral reef environments, adventure and sport. It was also widely believed that the Reef was 'neglected'. This was not intended to mean the Great Barrier was a mistreated environment, but rather that it was a neglected commercial opportunity. In 1929, Australian author Sydney Elliot Napier (1870–1940), who had recently returned from a naturalists' tour of the Reef, was quoted in the press on the subject of the region's potential for development:

> If properly exploited, I believe we would have in Australia one of the greatest tourist attractions in the world,' said Mr S. Elliott Napier, lecturing on 'A Holiday on the Great Barrier Reef,' at the Royal Empire Society. 'The Governments responsible,' he continued, 'have had this asset in their control, and they have failed to make proper use of it ... From a commercial point of view, or from that of the tourist, its riches are practically untouched.[10]

The 1920s saw a push by public figures to encourage Australian government bodies to attract tourists to the Great Barrier Reef. By 1934, the commercial and environmental value of the Great Barrier Reef was widely accepted, yet the region continued to be described as having 'potentialities not yet exploited [among] Nature's most beautiful work'.[11] Deirdre Gilfedder notes the difficulty of early efforts to tempt international as well as national visitors to the Great Barrier Reef.[12] The idea of travelling far from the city and mainland to an ocean wilderness was exciting, but also terrifying. To increase the visibility of the Reef for national and international tourism and produce a positive image, Australia's most accomplished modern commercial artists were deployed for the design of travel posters. As John Urry argues, tourism is a culture that is sensitive to the way people consume places visually.[13] To assist the visual consumption of Great Barrier Reef, eye-catching posters and advertisements were commissioned by Australian organisations including the Queensland Government Tourist Bureau and the Australian National Travel Association (ANTA), a federal institution that marketed Australia from 1929.[14] They helped create the coral reefs as objects of desire and fantasy, and tried to keep at bay any feelings of fear and trepidation towards the oceanic unknown.

Nature as spectacle and the split-level view

A favourite subject in the 1930s for persuasive images of the Great Barrier Reef was submarine scenery. But this was at a time before self-contained breathing apparatus (scuba) enabled bodily viewpoints from below. In the 1930s, it was a challenge for graphic artists to visualise and depict a living coral reef from an underwater perspective. One answer lay in replicating the visual experiences of viewing a tropical aquarium, and in mediating the experience of tropical coral reefs through aquarium optics.

How might a two-dimensional tourist advertisement replicate the experience of looking through an aquarium? By making the picture plane of the advertisement duplicate the viewing portal of a tropical tank and by adopting a split-level viewpoint (Figure 1). Aquariums enable viewers to look in and through the underwater, but also above water level to air. The split-level viewpoint embraces this dual perspective. It accentuates the liminal threshold between a volumetric perspective of the underwater, and a planar view of the air space above the water's surface.

Split-level viewpoints have been linked to scientific illustrations as well as fine art. Geologist Henry De la Beche (1796–1855), for example, applied a split-level viewpoint to his celebrated 1830 watercolour, *Duria Antiquior — A More Ancient Dorset,* depicting the evolution of prehistoric life from the undersea to land.[15] However, the pictorial structure showing a dual viewpoint of underwater and air came to the fore in the later nineteenth century after the emergence of aquariums for science and their popularity for the display of animals in private and public spheres. The nineteenth-century American artist Winslow Homer, for example, applied the optical effect of aquarium vision to his paintings, including *The Mink Pond* (1891). Art historian Charles C Eldredge refers to the effect in Homer's painting as 'split focus'.[16]

Symbolic value

What symbolic value did the split-level viewpoint contribute to tourist advertisements of the Great Barrier Reef? Two images (Figures 1 and 2) – one from the 1930s and one from the 1960s – graphically illustrate a sustained interest over time in depicting a distinct separation between the realm of the human, shown above the water's surface, and the realm of nature, shown as the underwater. Rather than portraying what Kari Wiel terms 'the physical and sensate world we share with animals', the images draw a line between human and marine animal realms, a division that reflect the period's attitude to the ocean as alien, remote and curious.[17]

In Figure 1, a vibrant 1933 lithograph and travel poster by Percy Trompf (1902–64), commissioned by the Queensland Government Tourist Bureau for the globalising tourism industry, we see the view below presented close-up and magnified like the effect of viewing an underwater scene through glass.

The view above water, however, is presented as a distant viewpoint, a landscape perspective that appears to float above the underwater in a world of its own. Three people — possibly children — look down through the water's surface. Dressed in white, the uniform of Europeans in the tropics, their appearance marks the enduring link between colonial explorers and a national identity framed by adventure in the outdoors.[18] Their white attire coupled with their quest to penetrate the water serves to remind the viewer about colonial explorers and their aspirations for knowledge, adventure and conquest. It is ambiguous whether they are naturalists with nets or fishers with lines and hooks, but there is a sense that their intention is to observe, trap and extract what lies below. One person has a waterscope, a glass optical device that when pushed through the surface of the water enlarges the view below. The waterscope utilises the material of glass to magnify the underwater and bring it closer to the eye. But while it allows a transparent view through water, it also acts as a barrier to the world beyond. The simultaneous evocation of closeness and remoteness, the separation of air from water by glass, the distance created between viewer from viewed all enable the waterscope to offer an experience of sublime wonder by keeping any potential danger at a distance.

The majority of the image in Figure 1 is below the water, where the wonders of the Great Barrier Reef lie in a combination of the bizarre and the beautiful. But the scene has the compressed aesthetics of an aquarium and the colours and clarity of an artificial undersea. The colourfulness of the scene does not reflect the optics of the undersea. As Shin Yamashiro explains, the colours of the undersea cannot be captured clearly by the eye because 'the depth of water influences how much sunlight can penetrate into the water: the deeper, the less colorful'.[19] Yet in Trompf's image, the saturation of colour is evenly distributed from land and air above to the sea floor below, underscoring the reality that natural vision and nature itself are not references for the image. Instead, Trompf has ensured that the non-human domain of the underwater is seen through the same optical laws, the same models of vision, as humans looking through air.

There is also the inference that the people in the boat cannot see what lies beneath them, whereas the viewer has been given a special look, as if the curtains have been drawn back just for them. It is a spectatorial device also found in painting and cinema. The image in Figure 1 therefore invites the narcissistic suggestion that nature has been designed for people, a point that relates to Timothy Morton's argument that the history of 'Nature' often entails the reduction of 'nonhuman beings to their aesthetic appearance for humans'.[20] The projection of a desired aesthetic on the coral animals of the undersea in Figure 1 is expressed through the artist's attention to stripes, tassels, blossoms and filaments, illuminated by shafts of sunlight through the water surface. It produces an orientalist effect in the sense of a generalised Western impression of a non-Western exotic.

It was common in the 1930s to compare a living coral reef to Western ideals of oriental aesthetics. According to the journalist E. W. Bogner, writing in 1933, the Queensland Government Tourist Bureau had prepared a booklet for worldwide distribution that compared corals and marine animals at the Great Barrier Reef to the riches of Aladdin's Cave and the 'raiment of an oriental prince': exotic, colourful, mysterious and with an element of danger.[21] The same booklet compared coral polyps to 'hoary Egypt's ancient architects', who built the pyramids.[22] The scale of the Great Barrier Reef and the pyramids were seen as mutually sublime, while their detailed forms and the myriad colours of corals were recognised as the epitome of beauty and were compared with the arabesque patterns of an oriental carpet. Gísli Pálsson argues that environmental Orientalism is part of the same 'colonial regime' of European mastery and domination over nature and exploitation based in the separation of society and nature.[23]

Humans as masters of nature are invoked by the image in Figure 2. Tourists at the Great Barrier Reef became practised at removing exotic animals from the reef for ornaments and souvenirs. Ben Daley and Peter Griggs explain how the Reef was 'loved to death' by travellers who believed it unthinkable to journey there and not return with curios of nature from one of the wonders of the world.[24] Starfish, for example, were among the most commonly collected animals from Great Barrier Reef. They were caught and boiled, and transformed from deep-blue living creatures to deep-red inanimate ornaments.[25] In Figure 2, a reef fossicker removes a starfish from the reef floor at low tide and turns it into an ornament by symbolically displacing it from its own milieu of water to the human realm of air. As a visualisation of tourism at the Great Barrier Reef, the image raises a question about the part played by images in naturalising the idea of human entitlements to master and control the animals of the Reef, including starfish and corals.

The coral realm beneath the surface of the water in Figure 2, which comes into focus through the visual trick of the split-level viewpoint, is not only shown as separate from the woman who stands on the reef but is also visualised by the artist as hers to take, and hers to rearrange. The image lends clear support to an argument forwarded by Aleksandra Ilicheva who wrote that 'our knowledge of corals comes from destruction and our desire to know and experience corals leads to destruction'.[26] The human figure is like a puppeteer with a stage of objects at her disposal. The person who looms large in Figure 2 appears to walk on water to survey the world below from a god-like viewpoint that can be compared with the 'god-trick' named by Donna Haraway as 'a perverse capacity … to distance the knowing subject from everybody and everything in the interests of unfettered power'.[27] The image was designed to promote human wonder in the presence of wilderness, but it also constructs an image of power. It does this through the gigantic scale of the human figure, but also by framing the non-human realm through the split-level viewpoint where the sea is domesticated by an aquarium aesthetic. Figure 2 allows the viewer to visualise the intersection of oceanic and terrestrial spheres. It promises an ecological viewpoint by showing the environmental context. However, it functions to tame the fear of the alterity of the underwater by reinforcing a reassuring divide between human and non-human realms. It promotes the kinds of human activities that soon led to the Reef's deterioration, and that was protested by Judith Wright and the Wildlife Preservation Society of Queensland when they became political in 1963 in their opposition to 'coral collectors, shell collectors and tourist interference'.[28]

Is there any evidence that suggests that the desire for power over the wilderness of the Great Barrier Reef was related to the emotion of fear? The proof exists in archival newspaper stories. Frank Hurley, the Australian explorer and photographer who travelled the length of the Great Barrier Reef in 1921, proposed that the zone beneath the surface of the sea represents 'the fear of the unknown'.[29] It was common in the 1930s to talk about the hidden 'monsters' of the Barrier Reef, especially in relation to large fish — including sharks and manta rays — and in one article the Reef was characterised as a place that has 'gorgeous colour — but where death is lurking'.[30]

Fear and the oceanic unknown

The geography of the Great Barrier Reef was a reminder that the uniqueness of the region is mostly under water. As a network of coral reefs, mostly submerged, the Reef is primarily an invisible marine space in three dimensions below the surface of the sea. This is where its psychological power resides. In *Water and Dreams*, Gaston Bachelard wrote in 1942 about the psychological difference between being 'on' and being 'in' water.[31] Being in the underwater means being placed where humans cannot breathe (without technologies) and into the natural element of marine animals, including predatory animals, as well as uncanny and unfamiliar undersea rock forms, and sinuous, mobile and entwining plants such as seaweed.

For tourists in the 1930s who considered making the journey to the remote islands of the Great Barrier Reef, the region aroused the dual emotions of attraction to beauty, but fear of the invisible and the unknown. A popular way of trying to engage with the underwater was with glass-bottomed boats. The more popular islands, including Green Island, built their own aquariums to attract tourists.[32] They enabled tourists to safely view the live animals of the Reef and encounter them eye-to-eye rather than from a viewpoint above. Bernd Brunner has argued that aquariums are a way of taming the animals of the undersea world 'by compressing them into an easily comprehended menagerie, an oceanic garden in miniature, a submarine chamber of wonders'.[33] A different way of utilising visual representations to attract tourists to the Reef, but also to alleviate any fear of the dangers of invisible forces of nature in the underwater, was to advertise the pleasure of snaring, trapping, hooking and killing large marine animals (Figures 3 and 4).

Overexposure and the animal body

The Great Barrier Reef in the 1930s was said to be teeming with marine life, and tourist organisations saw this as a significant opportunity for commercial nature tourism based in hunting and fishing. The large fish that inhabited deeper water were a special drawcard. As Figures 3 and 4 indicate, among the region's many attractions in the 1930s were game fish, especially black marlin, Spanish mackerel and sharks.[34] It was popular to design tourist posters and advertising images that intimated that an antidote to the anxieties of modern life was to control the wilderness. Typically, the images show large game fish hooked and struggling in mid-air for freedom in order to emphasize the fisherman's moment of triumph.

Paradoxically, in the first decades of the twentieth century, the idea of positive interactions with coral reef animals was often connected with their death, exploitation and mistreatment:

> The immense variety of fish offers a super-paradise to anglers. Turtle-riding excites endless hilarity. Crab-hunting on reef and beach entails tremendous fun. Spearing of eels, surfing, searching for eggs, shooting among countless birds of almost innumerable sorts and so forth.[35]

Australian zoologist T.C. Roughley (1888–1961), who was also president of the Great Barrier Reef Game Fish Angling Club (1937), enticed anglers to the Great Barrier Reef in 1937 with the promise 'we scarcely yet know what great fish these waters contain'.[36] Deep sea anglers were promised 'fish that are there for the catching, and they are a real fisherman's paradise'.[37] In the 1930s, the Great Barrier Reef was described as 'unspoiled by man'.[38] But in reality, by 1923 the Reef was already being fished out.[39] Today sharks and other species of the Reef's largest fish face ecological extinction from human impact.[40] Former chief scientist with the Australian Institute of Marine Science, Charlie Veron, remembers that it was way back in the 1960s that he last saw sharks 'by the dozens if not hundreds during every dive on any outer reef'.[41] The fear today is that those days of a teeming reef have gone and will never return.

James Northfield (1887–1973) was commissioned in the 1930s by the Australian National Travel Association to design posters promoting the Great Barrier Reef as a place of teeming wildlife. Through the split-level viewpoint of Figure 3 we see a viewpoint below the water's surface to the realm of fish and marine life and a viewpoint above where human presence dominates. The water's surface divides the world into human and other, and creates an asymmetry in power oriented to the opposition of the subject who hunts and the object that is hunted. For a tourist in 1932, Northfield's image might have signified the triumph of man, the thrill of the hunt and the control of nature. As a posthuman narrative, however, the isolated animal in Figure 3 reads as a symbol of the colonisation of nature and containment of its 'monsters', as does Figure 4. The thrill of the hunt, the magic of fishing, the conquest of nature, the pursuit of the monster and human bravado are all signified by Figure 4. The light of the moon allows the fight between man and beast to be silhouetted against the sky. Caught is the sublime moment that the marlin launches itself from the underwater into the air, transgressing the surface of the sea, which traditionally is the border between human and animal, yet safely kept in place and at a distance by a pictorial device that isolates it like an object and by the silhouette of a line and hook that it struggles against.

In the 1930s, there were no inhibitions about the desire to exploit the riches of the reef, especially fish for tourism, sport and industry, or to exploit the resources of the Reef to attract international tourists and capital. In 1936 — a time before tagging and before catch and release practices for fishers — the ocean was conceived as an aquarium filled with fish and marine animals for human amusement and sport. The waters of the Great Barrier Reef seemed a limitless resource. However, in 2009 in *The World is Blue,* Sylvia A. Earle observed that, 'The vision of a limitless ocean mesmerized policymakers, encouraging practices that have accelerated the depletion of marine wildlife and minerals: destroyed irreplaceable ocean species and ecosystems.'[42]

Final reflection

Wilderness has traditionally meant emptiness. The paradox of modernist tourist images of the wilderness of the Great Barrier Reef is the inclusion of human figures. They are included to help the viewer imagine themselves in the romantic position of being alone in the outdoors. Wilderness, however, is a colonial fantasy, as William Cronon shows. As he explains, the fantasy of virgin space is related to the colonial regime. In the United States, it has been responsible for the removal of Native Americans from sites deemed wilderness so that 'tourists could safely enjoy the illusion that they were seeing their nation in its pristine, original state, in the new morning of God's own creation'.[43] The story is similar in Australian history.

In 1927, it was claimed about the Great Barrier Reef that the 'islands and reefs have no population'.[44] Indigenous peoples, their histories, traditions and ownership were denied. Indigenous peoples are absent from the tourist images of the Great Barrier Reef in this article for similar reasons to those outlined by William Cronon. In order to uphold the fantasy that the Reef was an untouched wilderness, Indigenous owners who had lived, navigated and fished there for thousands of years were removed by whites to mainland coastal sites.[45] In fact, as Bernhard Klein argues, it is no coincidence that aquariums — and the views of capture and containment that they engender — emerged in Western countries during colonial times. There is a significant overlap between the aquarium's aestheticisation and containment of marine life and racial oppression, including the subordination and containment of Indigenous maritime peoples.[46] Early twentieth-century tourist lithographic posters of the Great Barrier Reef reveal the impact of aquariums on the visual consumption of the Great Barrier Reef. Their purpose was to display the wonders of the Great Barrier Reef, but they embody an extractive mentality relating to a will to exploit the Reef.

However, the invention of underwater cameras and scuba after World War II allowed tourists at the Great Barrier Reef to 'linger under water, suffused in the submarine spectrum'.[47] Scuba allowed interspecies encounters through bodies, and 'corporeal compassion', to use Ralph Acampora's phrase. He also writes that new technologies enabled interspecies encounters through 'animate bodies — bodies that are experienced and come to be known through interaction with other animate bodies'.[48] Such interactions have been characterised by Astrida Neimanis as a 'kinship imaginary'. With reference to the philosophies of Alphonso Lingis, Hélène Cixous and Catherine Clément — all of whom connect the human body with the sea — Neimanis also describes the human body as a watery echo of 'our fishy, watery beginnings'.[49] Lingis even imagined 'the plankton and krill, sponges and gorgonians' within his own body, thereby conceiving of himself as having 'inner coral reefs'.[50] In a postmodern, posthuman society, the bodily orientations of phenomenology and new technologies that enable human bodies to immerse themselves in the sea offer a new freedom for thinking of the human being as a planetary creature entangled with rather than remote from non-human life. But artists deployed by the tourist industry to promote the Great Barrier Reef in the 1930s utilised an aquarium viewpoint to domesticate and contain an otherwise fearful and unknowable oceanic realm.

Notes

1 See Ann Elias, *Coral empire: Underwater oceans, colonial tropics, visual modernity* (Durham, NC: Duke University Press, 2019).

2 Frank Hurley, 'The wondrous sea floor', *The Richmond River Herald and Northern Districts Advertiser*, 11 February 1921, 3.

3 Celmara Pocock, 'Romancing the Reef: History, heritage and the hyper-real', unpublished PhD thesis, James Cook University (2003), p. 240.

4 See Michelle Hetherington, *James Northfield and the art of selling Australia* (Canberra: National Library of Australia, 2006).

5 Jane Bennett, *Vibrant matter: A political ecology of things* (Durham, NC: Duke University Press, 2010), p. 116.

6 W. J. T. Mitchell, 'Showing seeing: A critique of visual culture', *Journal of Visual Culture* 1(2) (2002), 170–1.

7 Randy Malamud, *An introduction to animals and visual culture* (Basingstoke: Palgrave Macmillan, 2012), p. 3.

8 Judith Hamera, *Parlor ponds: The cultural work of the American home aquarium, 1850–1970* (Ann Arbor, MI: University of Michigan Press, 2012), p. 9.

9 Aquariums were nineteenth-century optical technologies that innovated ways of seeing and conceptualising the undersea and its animals. See Palle B. Petterson, *Cameras into the wild: A history of early wildlife and expedition filmmaking 1895–1928* (Jefferson, NC: McFarland & Co., 2011), p. 55.

10 Anon, 'Great Barrier Reef', *Sydney Morning Herald*, 20 June 1929, 15.

11 Anon, 'Wonders of Great Barrier Reef. Tourist asset to Queensland. Potentialities not yet exploited', *Nambour Chronicle and North Coast Advertiser*, 26 January 1934, 4.

12 Deirdre Gilfedder, 'The visual rhetoric of Australian travel posters between the wars: Branding a new nation', *Cultures of the Commonwealth* 15–16 (2010), 99.

13 John Urry, *Consuming places* (London: Routledge, 1995), p. 1.

14 Gilfedder, 'The visual rhetoric of Australian travel posters', 97.

15 For a discussion of De la Beche see Renee M. Clary and James J. Wandersee, 'Through the looking glass: The history of aquarium views and their potential to improve learning in science classrooms', *Science & Education* 14 (2005), 583. For an illustration, see Tom Sharpe, 'The De la Beche archive at Amgueddfa Cymru', National Museum Wales, https://museum.wales/articles/2009-04-20/The-De-la-Beche-archive-at-Amgueddfa-Cymru.

16 Charles C Eldredge, 'Wet paint: Herman Melville, Elihu Vedder, and artists undersea', *American Art* 11(2) (1997), 113.

17 Kari Weil, *Thinking animals: Why animal studies now?* (New York: Columbia University Press, 2012), p. 35.

18 National identity informed by narratives of early explorers is discussed in Gordon Waitt, 'Selling paradise and adventure: Representations of landscape in the tourist advertising of Australia', *Australian Geographical Studies* 35(1) (1997), 47–60.

19 Shin Yamashiro, *American sea literature: Seascapes, beach narratives, and underwater explorations* (New York: Palgrave Macmillan, 2014), p. 95.

20 Morton, 'X-ray', in Jeffrey Cohen (ed.), *Prismatic ecology: Ecotheory beyond green* (Minneapolis, MN: University of Minnesota Press, 2013), p. 311.

21 *The Great Barrier Coral Reef*, published by the Queensland Government Tourist Bureau, quoted in E. W. G. Bogner, 'The Great Barrier Coral Reef', *Horsham Times* (Vic), 14 July 1933, 10.

22 Bogner, 'The Great Barrier Coral Reef', 10.

23 Gísli Pálsson, 'Human–environmental relations: Orientalism, paternalism and communalism, in Philippe Descola and Gísli Pálsson (eds), *Nature and society: Anthropological perspectives* (London: Routledge, 1996), p. 67.

24 Ben Daley and Peter Griggs, '"Loved to death": Coral collecting in the Great Barrier Reef, Australia, 1770–1970', *Environment and History* 14(1) (2008), 97.

25 Anon, 'Wonders of Great Barrier Reef', 4.

26 Alexsandra Ilicheva, 'Plenty of fish in the sea: Geography of abundance, sustainability and ethics of non-mammalian marine animals', unpublished MA thesis, California State University (2011), p. 45.

27 Donna Haraway, 'Situated knowledges: The science question in feminism and the privilege of partial perspective', *Feminist Studies* 14(3) (1988), 581.

28 Judith Wright, *The coral battleground* (Melbourne: Spinifex Press, 2014), p. 2.

29 Frank Hurley, 'Beneath the waves: strange grotesque life; forests of amazing foliage', *Sun*, 11 November 1921, 9.

30 Norman Caldwell, 'The fangs of the Australian sea', *Voice* (Hobart), 5 October 1935, 2.

31 Gaston Bachelard, *Water and dreams* (Dallas, TX: Pegasus Foundation, 1983), p. 179.

32 N.Q. Naturalists' Club, 'Green Island Aquarium', *Cairns Post*, 4 September 1936, 3.

33 Bernd Brunner, *The ocean at home: An illustrated history of the aquarium* (New York: Princeton Architectural Press, 2005), p. 126.

34 Anon, 'Thrills and joys of the Barrier Reef: Teeming with game fish', *Proserpine Guardian*, 24 August 1935, 3.

35 Bogner, 'The Great Barrier Coral Reef', 10.

36 T. C. Roughley, *Big game angling Australia* (Sydney: National Travel Association, 1937), unpaginated.

37 Anon, 'Wonders of Great Barrier Reef', 4.

38 Anon, 'Thrills and Joys of the Barrier Reef', 3.

39 H. G. McKay, 'Undeveloped resources of the Barrier Reef', *Smith's Weekly*, 1 September 1923, 4.

40 William D. Robbins, Mizue Hisano, Sean R. Connolly and J. J. Howard Choat, 'Ongoing collapse of coral-reef shark populations', *Current Biology* 16 (2006), 2314.

41 J. E. N. Veron, A reef in time: The state of the Great Barrier Reef (Cambridge, MA: The Belknap Press of Harvard University Press, 2008), p. 49.

42 Earle cited in Will Abberley, 'Introduction' in *Underwater worlds: Submerged visions in science and culture* (Newcastle: Cambridge Scholars Publishing, 2018), p. 10.

43 William Cronon, 'The trouble with wilderness: Or, getting back to the wrong nature', *Environmental History* 1(1) (1996), 15.

44 Anon, 'The Great Barrier Reef: Nature's bounty', *Port Macquarie News and Hastings River Advocate*, 10 December 1927, 4.

45 For a discussion of colonial impacts and dispossession at the Great Barrier Reef see I. Lyons, R. Hill, S. Deshong et al. 'Putting uncertainty under the cultural lens of Traditional Owners from the Great Barrier Reef catchments', *Regional Environmental Change* 19 (2019), 1597–1610. For information about Indigenous ownership and historical displacement, see Great Barrier Reef Marine Park Authority, 'Traditional Owners of the Great Barrier Reef', https://www.gbrmpa.gov.au/our-partners/traditional-owners/traditional-owners-of-the-great-barrier-reef.

46 Bernard Klein, 'The ocean at home: An illustrated history of the aquarium' review in *Victorian Studies* 48(4) (2006), 710.

47 J. Malcolm Shick, 'Towards an aesthetic marine biology', *Art Journal* 67(4) (2008), 78.

48 Ralph R. Acampora, *Corporal compassion: Animal ethics and philosophy of body* (Pittsburgh, PA: University of Pittsburgh Press, 2006), p. 5.

49 Astrida Neimanis, *Bodies of water: Posthuman feminist phenomenology* (London: Bloomsbury, 2017), pp. 137–8.

50 Alphonso Lingis, 'The rapture of the deep', in Alphonso Lingis, *Excesses: Eros and culture* (New York: State University of New York Press, 1983), p. 13.

Ann Elias is Professor of Art History and Visual Culture at the University of Sydney. Research interests include camouflage as a military, social and aesthetic phenomenon; flowers and their cultural history; and coral reef imagery of the underwater realm. Books include *Camouflage Australia: Art, Nature, Science and War* (2011), *Useless Beauty: Flowers and Australian Art* (2015) and *Coral Empire* (2019). She is a Research Affiliate with the Sydney Environment Institute. Research in progress asks how the underwater of Sydney Harbour relates to Australian, Pacific and world social histories.

Basket case!

Carden C. Wallace
carden.wallace@qm.qld.gov.au

Figure 1.
A new basket sits on the Ribbon Reefs, Great Barrier Reef Australia, circa 2002.
Image P. R. Muir

Queensland has some 400 public museums and art galleries.[1] Large or small, these are all dedicated to caring for their part of what is often called the 'distributed national collection' and to permanently documenting a segment of our history — social, natural or otherwise.[2] Each of us who steps inside such an institution to help in this effort is liable to become lost to this world for the rest of our working life. We are all, in some sense, collectors, and we tend to be very loyal to 'our' subject matter.

My particular entrapment has been with the Queensland Museum — a place where the collections range from geology, natural history and human endeavours to books and documents of all kinds. I did not spend my whole career in this place, but it certainly played a large part. My obsession has been with the hard corals that form the backbone of the Great Barrier Reef. What's not to enjoy about these achingly beautiful underwater animals, their subtleties of physical difference, their roles in the reef ecosystem and the way they protect our coastline and that of

Figure 2.
J. Wolstenholme carries the basket in Indonesian waters, 1994.

tropical and subtropical parts of the world? They also have something to tell us about most of the questions in our changing times[3] and the dangers they face from global events.[4]

In science, the methodology is to hypothesise, test and challenge until a current question is answered, at least up to a new level of revelation and within the boundaries of the available technologies. As more and more researchers become involved, and new questions and issues emerge, a solitary curator begins a range of collaborations. Publications ensue, and suddenly the subject area has jumped state and national boundaries and applies to reefs all around the world. Perhaps, as was the case for me, this may force us to narrow down the range of questions, and compare answers from numerous locations and circumstances.

In short, I became a world traveller and international diver. My goal was to document the staghorn reef corals of the world, allowing latitude and longitude, depth of occurrence, diversity and abundance to be recorded and analysed in various ways, and to keep the evidence. The most interesting overseas place, for about ten years' duration, was Indonesia.[5] This vast land of islands sits between the two big oceans, Pacific and Indian, and represents a complexity of tectonic histories, movements of sea floor plates, and opening and closing of land barriers, all of which have been involved in the formation of this richest coral area in the world. Relatively little work had been published about the coral species composition of this wondrous place, but it first required visits to experts and museums in the Netherlands, learning (some) Indonesian in order to work with institutions and counterparts there and planning a complex work plan that allowed for sampling the full range of reef types (including underwater volcanoes and volcanic ash slopes) over time, as well as understanding monsoon patterns, means of access and potential for finding new species.

Thinking back on this, as well as on visits to reefs from India, the Maldives, Oman and the Seychelles to Thailand, Brunei, China, New Guinea, Japan, New Caledonia, French Polynesia, various islands across the central Pacific and even a little of the Caribbean, I realise that once I ventured into the broader world, logistics and planning became very important. All these places have scientific institutions, field laboratories, fisheries departments (these are usually very involved with coral reef studies), scientists and students, as well as boats, dive teams and researchers willing to help and share their knowledge. There was always plenty of company and assistance on offer, as well as the camaraderie of like-minded people (though of different nationalities). When flying off to these places, the scientific luggage — including dive equipment and cameras — was always more important than personal gear and clothes: fortunately, reefs are mostly in warm places. The only problem was that the fieldwork required quite a lot of simple but heavy equipment, which just could not be carried around the world – hammers, chisels, buckets, containers and chemicals of various kinds. (These are all things that you just 'throw in the truck' for fieldwork at home.) Oh, and most important of all – a basket! These things had to be acquired in the destination country.

The standard Queenslander's basket for underwater use was of the kind used in supermarkets – sturdy plastic, with a black handle. I first encountered it when working with the legendary John 'Charlie' Veron on the Great Barrier Reef.[6] In almost forty years of overseas diving, I must have purchased about 80 baskets at roadside stalls in various places. Once underwater, I'm pleased to be able to say that I collected coral samples as one would pick flowers, so no corals were killed in the process. The importance of the basket cannot be overstated. The corals had to be recorded on an underwater slate by colour, depth and other characteristics, and a numbered label attached. Laden with a few samples, the basket would not float away and it could be rested on the reef as needed. When the basket was full, it was taken up to the dive boat and the corals would later be bleached, washed and dried in the sun. The accumulated specimens, once written up in a notebook, would eventually be carefully wrapped and packed for sending home via a freighting company. It was very important to have a Convention on International Trade in Endangered Species (CITES) permit, as corals are highly protected.[7]

When the fieldwork was over, the basket, along with the other tools purchased in country, would be passed on to local colleagues. I wish I had a way of honouring this sturdy workhorse, which has accompanied me, in many forms and colours, on so many journeys under the sea. It has never let me down.

Notes

1 See http://www.magsq.com.au/museum/finder.asp.
2 Margaret Henty, 'The Distributed National Collection', *Australian Academic & Research Libraries* 22(4) (1991), 53–9; Andrew Simpson, 'Cinderella, fifteen years after the ball: Australia's university museums reviewed', *Museums Australia Magazine,* 21(2) (2012), 18–20.

3 P. R. Muir, C. C. Wallace, T. Done and J. D. Aguirre, 'Limited scope for latitudinal extension of reef corals', *Science* 348 (2015), 1135–8; C. H. White, D. W. J. Bosence, B. R. Rosen and C. C. Wallace, 'Response of Acropora to warm climates: Lessons from the geological past', in *Proceedings of the 11th International Coral Reef Symposium, Ft. Lauderdale, Florida, 7–11 July 2008*, Vol. 1 (2010), 7–12.

4 W. J. Skirving et al., 'The relentless march of mass coral bleaching: A global perspective of changing heat stress', *Coral Reefs* 38 (2019), 547–57.

5 C. C. Wallace and J. Wolstenholme, 'Revision of the coral genus Acropora (Scleractinia: Astrocoeniina: Acroporidae) in Indonesia', *Zoological Journal of the Linnean Society* 123 (1998), 199–384.

6 J. E. N. Veron and C. C. Wallace, *Scleractinia of Eastern Australia. Part V. Family Acroporidae*, Australian Institute of Marine Science Monograph Series 6 (Canberra: ANU Press, 1984).

7 See https://cites.org/eng.

Carden Wallace AM is an Australian scientist who was the curator/director of the Museum of Tropical Queensland from 1987 to 2003. She is an expert on corals, having written a 'revision of the Genus Acropora'. Wallace was part of a team that discovered mass spawning of coral in 1984. She is Emeritus Principal Scientist at the Queensland Museum.

Great Barrier Reef World Heritage: Nature in danger

Celmara Pocock
celmara.pocock@usq.edu.au

Abstract

The Great Barrier Reef is inscribed on the World Heritage List for its natural values, including an abundance of marine life and extraordinary aesthetic qualities. These and the enormous scale of the Reef make it unique and a place of 'Outstanding Universal Value'. In the twentieth century, protection of the Great Barrier Reef shifted from limiting mechanical and physical impacts on coral reefs to managing agricultural runoff from adjacent mainland to minimise environmental impacts. By the early twenty-first century, it was apparent that threats to the Great Barrier Reef were no longer a local issue. Global warming, more frequent extreme weather events and increased ocean temperatures have destroyed vast swathes of coral reefs. Conservation scientists have begun trialling radical new methods of reseeding areas of bleached coral and creating more resilient coral species. The future of the Great Barrier Reef may depend on genetically engineered corals, and reefs that are seeded, weeded and cultured. This article asks whether the Great Barrier Reef can remain a natural World Heritage site or whether it might become World Heritage in Danger as its naturalness is questioned.

Introduction

The Great Barrier Reef is inscribed on the World Heritage List for its natural values, including an abundance of marine life and exceptional aesthetic qualities. These attributes, together with the enormous scale of the Reef, which stretches for some 2,300 kilometres along the north-east Queensland coast, distinguish it as unique and a place of 'Outstanding Universal Value'. Protection and management of the Reef is thus focused on the conservation of the listed natural attributes, which include aesthetic qualities but not landscape values.

In the twentieth century, protection of the Great Barrier Reef was focused largely on mitigating mechanical impacts and physical harm in localities impacted by activities such as coral collecting, reef walking and boat anchoring. These direct impacts could, to a large extent, be managed through local regulation and efforts to change human behaviours at the Reef. By the second part of the twentieth century, the widespread and devastating impacts of crown-of-thorns starfish outbreaks and coral bleaching had become major threats. Even though these too were human-induced changes, unlike the local mechanical damage, they had their origins outside the boundaries of the Great Barrier Reef World Heritage Area. The Great Barrier Reef Marine Park Authority (GBRMPA) therefore expanded its conservation measures from a focus on fishers and tourists to consider local communities on the adjacent mainland. However, by the early twentieth-first century, it was apparent that the threats to the Great Barrier Reef were even more diverse and could not be managed by simply altering local behaviours. Global issues of marine pollution and climate change wreak extensive damage on the Great Barrier Reef, undermining its integrity and status as a World Heritage property.

Global warming, an increasing number of extreme weather events such as cyclones, floods and fires, and the constantly warming oceans have had a dramatic impact on coral reefs globally, including coral bleaching of large tracts of the Great Barrier Reef. While these events and impacts are occurring at alarming rates, arresting the root cause remains a considerable political and practical challenge. As global leaders and environmentalists struggle to rein in climate change, scientists at the Reef have turned to a number of radical interventions to try to create a more resilient Great Barrier Reef. In the face of irreversible climate change, Reef management no longer seeks to change human behaviour, but to change the Reef itself. This article considers the implications of such interventions for the status of the Great Barrier Reef as a World Heritage site of *natural* significance.

World Heritage Listing of natural values

The Great Barrier Reef was inscribed on the World Heritage List in 1981, seven years after the Australian Government became a signatory to the UNESCO World Heritage Convention. At the time, World Heritage assessments were made under either cultural or natural nominations, and the Reef was listed under all four of the then natural criteria.

Since that time, the World Heritage criteria have been updated on several occasions as the system responded to emerging issues and attempted to reflect a broader range of approaches and expectations about how heritage should be identified and managed. A continuing challenge for World Heritage derives from its historic and ingrained separation of natural and cultural values, including the provision of two separate lists of criteria for inscribing natural and cultural World Heritage properties (Burke and Smith, 2010; Harrison, 2015; Lee, 2016; Lowenthal, 2005). This separation created a bias in the World Heritage List, with places of cultural significance almost entirely represented by European monuments and grand buildings, while natural properties were characterised as pristine, untouched wilderness (Smith, 2013). This dualism revealed that the World Heritage List was not genuinely representative of the heritage of all humanity. Notably, properties inscribed under 'natural' heritage ignored the ways natural areas are frequently a product of human perception, management and intervention. This is particularly problematic for Indigenous world-views, which regard nature and culture as indivisible (Lee, 2016; Lilley and Pocock, 2018). Efforts to redress the division between nature and culture in the World Heritage system include the introduction of a criterion enabling the recognition of cultural landscapes and the integration of World Heritage criteria into a single list.

The introduction of the cultural landscape criterion in 1992 aimed to capture a broader range of heritage properties, and the Operational Guidelines for the

Table 1. World Heritage criteria for which the GBR is listed

(vii)	to contain superlative natural phenomena or areas of exceptional natural beauty and aesthetic importance.
(viii)	to be outstanding examples representing major stages of earth's history, including the record of life, significant on-going geological processes in the development of landforms, or significant geomorphic or physiographic features.
(ix)	to be outstanding examples representing significant on-going ecological and biological processes in the evolution and development of terrestrial, fresh water, coastal and marine ecosystems and communities of plants and animals.
(x)	to contain the most important and significant natural habitats for in-situ conservation of biological diversity, including those containing threatened species of outstanding universal value from the point of view of science or conservation.

Source: UNESCO World Heritage Centre (2019).

Implementation of the World Heritage Convention encouraged nominations of mixed natural and cultural values (Aplin, 2007; Brown, 2015; Brown, 2012; Cleere, 1995; Rössler, 2002, 2006; Smith, 2013). To a large extent, World Heritage Cultural Landscapes focus on the capacity of this criterion to capture associated values — the types of values that more recently would usually be termed social or intangible values. As Smith (2013) argues, the landscape criterion has had limited impact on integrating natural and cultural values, and it is primarily through associative values that cultural landscapes have most effectively recognised this integration. One of the systemic issues that perpetuates this issue is how the landscape criterion remains embedded in the separation of nature and culture. The Operational Guidelines define cultural landscapes as follows:

> Cultural landscapes are cultural properties and represent the 'combined works of nature and of man' designated in Article 1 of the Convention. They are illustrative of the evolution of human society and settlement over time, under the influence of the physical constraints and/or opportunities presented by their natural environment and of successive social, economic and cultural forces, both external and internal. (UNESCO World Heritage Centre, 2019, p. 20)

While the separate criteria for natural and cultural listings were combined into a single set of criteria after 2004, the Operational Guidelines make it clear that Article 1 defines cultural heritage, and this includes cultural landscapes, whereas Article 2 defines natural heritage, which makes no reference to landscape or associative values. Further, while the criteria are now listed together, there is still a clear distinction between the first six criteria that relate to cultural heritage and the last four that pertain to natural values. It is thus easy to map the criteria used for earlier listings across to the new single list. For the Great Barrier Reef, which was listed under all four natural criteria in the earlier system, these relate directly to the single list of criteria outlined in Table 1.

As I have argued elsewhere (Pocock, 2002, 2020, 2022), criterion (vii) is capable of reflecting associative values for the Great Barrier Reef, and thus constitutes a form of cultural heritage, but the continued division of cultural landscapes as a category of cultural heritage denies this possibility. Instead, the Great Barrier Reef, a large-scale landscape and seascape, is definitively recognised as a purely natural heritage site under the World Heritage Convention. While this has long been a problematic categorisation, responses to the environmental crises facing the Great Barrier Reef in the twenty-first century exacerbate doubts about whether the categorisation of the Reef as natural heritage is sustainable.

The cultural values of Great Barrier Reef aesthetics

Many aspects of the Reef that are presumed to be natural are in fact produced by human knowledge and skill. While there is some superficial acknowledgement of Indigenous attachments to the Great Barrier Reef (McIntyre-Tamwoy, 2004, p. 23), there is almost no understanding of how the Reef might be considered cultural within non-Indigenous perceptions. Understanding the region as a single phenomenon, for instance, is a product of human experience and imagination, made palpable through mapping, aerial photography and satellite imagery (Pocock, 2004, 2006, 2009). These technologies are all culturally produced. Similarly beneath the surface, the astounding richness of underwater life — the diversity of brilliant corals, sponges and the abundance of colourful fishes, turtles and other life forms — is made accessible and communicable to others through technology.

A singular Great Barrier Reef

Initial European accounts of the Reef focused on the navigational dangers it posed to fragile timber ships. Their perception of the Great Barrier Reef came about through the charting and map-making to avoid collision with the myriad shoals and reefs. Contemporary understandings of the complexity and dangers of the Reef remained entwined with early colonial histories. This includes the voyage of the *Endeavour*, in which James Cook came perilously close to catastrophe when the ship collided with the now eponymous Endeavour Reef in 1770. But it was Matthew Flinders, some 30 years later, who realized that the region comprised a multitude of 'great barrier reefs' and named it as such. The plurality of this naming, however, tends to be overlooked in contemporary understandings of the Great Barrier Reef as a single entity. The idea of the Great Barrier Reef as a single phenomenon thus has origins in the deeply enculturated practices of European mapping that allow the viewer to see at scale a single entity gathered from a multitude of individual, diverse and complex phenomena.

Like maps, photographs are also cultural products, but the latter are more readily accepted as representations of an external reality. Aerial photography and satellite imagery often appear to give a reality to the maps created in an earlier era, and even inform and replace present-day mapping. These aerial images in turn produce aesthetically pleasing vistas that are synonymous with the expanse and beauty of the Great Barrier Reef. This is manifested in the way the Great Barrier Reef is recognised for its aesthetic qualities under natural criterion (vii). Thus photographic reproduction of the Reef as a single phenomenon is a powerful consolidation of public acceptance that the Great Barrier Reef is not a composition of multiple reefs, islands, shoals and regions, but a single observable entity. Photography further produces aesthetic qualities recognised as holding 'Outstanding Universal Value'. This is fundamental to the status of the Reef as a natural wonder of the world and as a World Heritage site of natural significance. The naturalness of this phenomenon is, however, only conceivable through the powerful cultural lenses of mapping and photography.

Underwater life

Photography has also produced the iconic imagery of the underwater that ignites the popular imagination and underscores the significance of Great Barrier Reef as a World Heritage property. Aboard the earliest European ships to navigate the Reef were a number of scientists who first glimpsed the colourful life of reefs from the decks of ships. Like all other resources of the new colonies, the waters of the Great Barrier Reef were originally studied for their economic potential and the work of the late nineteenth-century biologist William Saville-Kent remains a masterpiece of detailed study of the coral reefs.

A broader interest in the natural underwater world emerged as part of a naturalist tradition. Early excursions to the Reef were undertaken by ornithological groups from New South Wales and even South Australia, and by the beginning of the twentieth century the Australian Museum in Sydney had established a regular program of research at the Great Barrier Reef. Queensland naturalists, by comparison, were less involved, and research on the Reef was of negligible interest to the Queensland Museum. Rather, it was an expedition from Britain that highlighted the natural significance of the Great Barrier Reef when a group of scientists spent a year living on the Low Isles in 1928–29. There was great public interest in this expedition, and the British team was joined by scientists and journalists from Sydney, many associated with the Australian Museum. Descriptions of the wonders of the coral reefs were widely reported and published in newspapers, magazines and popular books far beyond scientific outlets, and this inspired many to follow (Pocock, 2010, 2020).

The Great Barrier Reef was a dangerous navigational obstacle, overwhelming in scale and difficult to reach. The model of Reef expeditions based on offshore islands was one of the only practical ways to experience the Reef at first hand. Boat voyages to the islands were very limited from the adjacent mainland, and the Outer Reef was all but inaccessible. The earliest holidaymakers thus accompanied Australian Museum expeditions to islands, staying with them for lengthy periods over the summer holidays. The most regular and celebrated of these expeditions were organised by a New South Wales school teacher, Mont Embury, and were led by both amateur and professional scientists through the 1920s and 1930s. Holidaymakers emulated scientific activities, learning from scientists, assisting in research and undertaking their own studies. Thus, both science and tourism developed through a single approach to understanding the Reef (Pocock, 2010, 2020). The cultural practice of tourism too became enmeshed in the science of nature, amplifying the Reef's natural heritage for lay audiences.

For all involved, access to the underwater held particular challenges. The most common way to view the living reefs was to wait patiently for the right tidal conditions. Visitors typically waited for the opportunity to peer into coral pools left by the receding tide, where from the surface they could observe fish and other life forms in a miniature form of the Reef. Waterscopes, also known as water glasses, could be used to view the underwater even when observation was obscured by deeper water or when the surface was rippled by wind or movement. These vignettes were a source of great delight and interest to scientists and holidaymakers alike. But studying life at closer quarters often required removing living creatures from their environment to where the dying or dead could be studied in greater detail.

These constraints remain for many aspects of scientific investigation, but more direct access to the underwater world through scuba diving equipment and underwater photography has greatly enhanced the research and study of living reefs. Photography in particular has been very influential in bringing the wonder and diversity of the coral reefs to a broad public (Pocock, 2009). Images of vibrant, colourful and extraordinary lifeforms are synonymous with the Great Barrier Reef. Together with the aerial images, these constitute the aesthetic significance of the Reef as a place of 'Outstanding Universal Value'.

Public access to these images was itself significant in garnering support for the Great Barrier Reef marine park and the eventual World Heritage Listing. Some of the most vociferous advocates for Reef conservation had only ever experienced the region through the extraordinary array of photographic imagery made possible by human technology, including underwater cameras, motion cameras, colour emulsion and increasingly sophisticated digital technology. Images are enhanced by filters, night photography and other techniques that seldom equate to the direct human vision underwater. They thus inspire action to conservation and shape management in the present.

Contemporary environmental challenges

Although arguments about the cultural nature of aesthetic appreciation might appear largely theoretical, the responses to contemporary environmental challenges make the cultural nature of the Reef manifest.

European Great Barrier Reef conservation concerns have existed for more than a century. At the beginning of the twentieth century, E. J. Banfield created a sanctuary on Dunk Island to protect birds from the wanton destruction of shooters (Banfield, 1908). By the mid-century, government had sanctioned threats to the Reef that emerged through proposals for sandmining and mineral exploration on offshore cays and reefs during the Sir Joh Bjelke-Petersen era (Daley and Griggs, 2006; Wright, 1977). Moreover, tourism inspired by a love of the Reef paradoxically also led to significant damage to the corals and reefs. Coral and shell collecting was a popular activity for much of the twentieth century, to the extent that some reefs were stripped bare (Daley and Griggs, 2008; Pocock, 2020). At the same time, walking across the reefs at low tide, whether to fossick or simply to look at the coral pools, killed the sensitive and fragile polyps that grow the coral. Reef walking and the anchoring of tour boats on the reefs continued until prohibited towards the end of the century. These forms of mechanical damage had significantly detrimental impacts on the coral, or the potential for damage, but to a large degree such threats were confined to particular localities. Consequently, solutions to these problems could be addressed through immediate action. Other impacts could be addressed by management policy and the implementation of policies that demanded change to localised human behaviours, such as prohibiting collecting and anchoring on reefs. Yet, while such localised damage could be remedied through relatively direct interventions, other problems emerged that had less immediate causes.

The idea that impacts on one part of the Reef could affect another part was central to the initial successful campaign to protect the Great Barrier Reef from mining and other impacts. In turn, the recognition of the Great Barrier Reef as a single phenomenon was key to the creation of the Great Barrier Reef Marine Park and subsequent World Heritage Listing (Bowen and Bowen, 2002; Wright, 1977). A key argument in the original battle to save the Great Barrier Reef drew on the then-emerging field of ecology. With its principles of holism and interconnectivity, a scientific argument could be made to understand the interconnection of many reefs and shoals as a single Great Barrier Reef (McCalman, 2017). Despite the importance of this argument for establishing the Great Barrier Reef Marine Park and the World Heritage Listing, an ecological approach to management only emerged much later in response to new kinds of problems (Olsson et al., 2008). Foremost among these were the outbreaks of crown-of-thorns starfish (COTS). A naturally occurring species on the Reef, COTS eat coral polyps as many Reef species eat one another. However, the conditions that led to outbreaks of plague proportions of COTS left a trail of destruction during a series of outbreaks beginning at Green Island in 1962 and 1979, with a third outbreak first detected near Lizard Island in 1993 (Miller et al., 2015). Attempts to manage these outbreaks initially followed established management approaches of localised intervention through physical removal of the invading starfish from heavily infested areas. However, these efforts appeared futile against a growing tide of COTS outbreaks. The discovery that COTS larvae thrive on phytoplankton, which increase in nutrient-rich waters (Fabricius et al., 2010) brought the realisation that human activity was a major contributor. Fertiliser runoff from agricultural businesses on the adjacent Australian mainland was shown to have a direct and negative impact on Reef health, even though it was ostensibly occurring some distance away. Management strategies thus had to shift from thinking about local Reef-based behaviours and solutions to engaging with communities further afield. A program of working with farmers to reduce runoff and fertiliser use has offered some benefits (Deane et al., 2018). However, cyclones and other events that bring increased sedimentation to the Reef are not so easy to control, and laborious manual eradication of COTS remains the most effective method of protecting reefs from outbreaks (Westcott et al., 2020). Sustaining the idea of distinctive Reef natural heritage flounders in the face of COTS outbreaks and other impacts that have their origin in diffuse and global human activity.

The longstanding conservation practice of creating protected areas to preserve biodiversity and other natural values is challenged globally by environmental issues that originate elsewhere but have widespread local impacts. These originate in behaviours that are both anonymous and belong to all of us. Issues such as plastic pollution, warming oceans, and increasing and more extreme climatic events including cyclones, floods and bushfires now impact many parts of the world. Such changes do not recognise the boundaries of national parks or marine conservation areas (Tweed, 2010). The impacts are widespread and devastating to the Great Barrier Reef World Heritage Area. The increased frequency and intensity of cyclones inflict mechanical damage on reefs and create broader conditions detrimental to Reef resilience, such as generating conditions that support significant increases in numbers of COTS. While reefs can recover between COTS outbreaks, this can take between ten and twenty years. The increased frequency of cyclones means that recovery becomes less and less likely as areas are repeatedly subjected to COTS threats without sufficient time to recover (Westcott et al., 2020). More broadly, warming oceans that lead to widespread coral bleaching are having a devastating impact on coral reefs globally, including the Great Barrier Reef.

World Heritage Committee reactive mission 2012

The extreme impact of climate change on the Great Barrier Reef led the World Heritage Committee to request an investigation in the form of a reactive monitoring mission to the Great Barrier Reef in 2012.

The mission found that earlier threats to the Reef were being well managed, including previously noted issues such as oil and gas development, recreation, fishing and tourism, and 'most recently water quality from catchment run-off', and that these were 'likely be further improved in the future'. While the mission prioritised water quality and port activity as the most immediate threats to the Outstanding Universal Values of the Great Barrier Reef, it secondarily acknowledged that climate changes, including increased numbers of cyclones, were having direct and indirect impacts on the Reef. However, despite finding that the World Heritage Area was affected by a number of current and potential threats, and that the environmental quality of some Reef ecosystem had declined since it was inscribed in 1981, the joint mission by the International Union for the Conservation of Nature (IUCN) and the International Committee on Monuments and Sites (ICOMOS) found that the Reef continued to retain its Outstanding Universal Value (Douvere & Badman, 2012). In other words, it found that the Great Barrier Reef remains a property of outstanding *natural* value.

While the IUCN and ICOMOS report emphasises reducing environment impacts to enhance the natural resilience of the Great Barrier Reef, there is a growing body of research suggesting that 'conventional management approaches will be insufficient to protect coral reefs, even if global warming is limited to 1.5 °C' and that 'emerging technologies are needed to stem the decline of these *natural* assets" (Anthony et al., 2017, emphasis added). There is thus a trend towards management interventions to support and enhance the resilience of reefs. Management of the Great Barrier Reef now includes some experimental methods, including seeding rain, reducing sunlight and engineering more heat-tolerant coral species. For instance, a chemical sunshade is being trialled to reduce the amount of sunlight penetrating the water. Increasingly, scientists are experimenting with using coral larvae collected during the annual coral spawning events to repopulate areas of the Great Barrier Reef that have been damaged by mass bleaching events (Great Barrier Reef Foundation, 2017; Suggett et al., 2019).

These new management activities are a long way from seeking simply to reduce or remove human threats. Rather, they actively intervene in biological processes to produce the range, tolerance and growth of corals in areas where it has become unsustainable due to temperature, water acidification and pollution. There is no doubt that the cultivation of more resilient species and similar interventions, such as providing structural supports, and growing, nurturing and replanting broken corals, are actually cultural activities.

Natural values in danger

The aesthetics of the Great Barrier Reef can thus be argued to be as cultural as much as natural. Indeed, photography and Reef science are so strongly intertwined that they are mutually constitutive (Pocock, 2009). The interconnections between scientific research, photographic technology and heritage aesthetics of the Great Barrier suggest that in reality the Reef is a nature-culture landscape. This conceptualisation might appear abstract to those who research, study and explore the Reef through a scientific tradition that regards the natural world as external, measurable and objective, and apart from human experience. Such a separation of humans from nature is at the heart of Western thinking and research, and it is embedded in the way the World Heritage system has assessed, and largely continues to assess, nature and culture by using distinct criteria. This dualism has, however, been challenged for creating imbalance and an unrepresentative World Heritage List, and the World Heritage Centre has responded by adapting and creating new criteria for assessing World Heritage. Despite these efforts, the integration of culture and nature is still largely regarded as an issue for the 'other' — for Indigenous and local custodial knowledge systems — rather than as a core reconsideration of World Heritage processes. However, the contemporary issues now facing the future of the Great Barrier Reef are bringing the cultural nature of the Great Barrier Reef to the fore.

While the joint mission by IUCN and ICOMOs found that the Reef continued to retain its Outstanding Universal Value, it also warned that future threats to the Reef would come from increased water temperatures, and more frequent and more intense weather events (Douvere & Badman, 2012). These are threats experienced by coral reefs globally. A special World Heritage Committee meeting on the resilience of coral reefs recognised that the only way to ensure coral reef survival is through limiting increases in global warming, an issue that can only be addressed on a global scale (UNESCO 2019). Despite this knowledge, conservation scientists to some extent accept that global warming is inevitable, and they have initiated radical new interventions to sustain the Reef, including attempts to create corals that can withstand the changes of our time and to reseed bleached corals. Their effectiveness in keeping the Reef alive is yet to be seen. However, if this is the only way the Great Barrier Reef can continue into the future, then its status as a place of outstanding natural value must surely be brought into question. While intended to conserve the Reef, new scientific approaches appear to undermine the very qualities that underpin its status as a natural phenomenon. Even the most elemental dualistic understanding of 'nature' and 'natural' is surely unsustainable as the Reef increasingly becomes a physical product of human labour. The World Heritage system continues to grapple with how to integrate cultural and natural values, largely in response to a need to integrate non-Western and Indigenous understandings of landscape into its processes. However, the future of the Great Barrier Reef appears to rest on cultural interventions as much as it does on biological processes, and recognising this integration may be critical to its World Heritage status in future. As technological advances become more necessary to counter global impacts that threaten to place the Reef on the list of World Heritage in Danger, its status as a *natural* property is already in danger.

Acknowledgements

I would like to thank the editors, Kerrie Foxwell-Norton and Iain McCalman, for inviting me to contribute to this special issue, and for their helpful suggestions on the manuscript.

References

Anthony, K., Bay, L. K., Costanza, R., Firn, J., Gunn, J., Harrison, P., ... Moore, T. 2017. 'New interventions are needed to save coral reefs', *Nature Ecology & Evolution* 1(10): 1420–22.

Aplin, G. 2007. 'World Heritage Cultural Landscapes', *International Journal of Heritage Studies* 13(6): 427–46.

Banfield, E. 1908. *The confessions of a beachcomber*. Sydney: Angus & Robertson.

Bowen, J. and Bowen, M. 2002. *The Great Barrier Reef: History, science, heritage*. Melbourne: Cambridge University Press.

Brown, J. 2015. 'Bringing together nature and culture: Integrating a landscape approach in protected areas — policy and practice. In R. Gambino and A. Peano (eds), *Nature policies and landscape policies: Towards an alliance*. Cham: Springer, pp. 33–42.

Brown, S. 2012. 'Applying a cultural landscape approach in park management: An Australian scheme', *Parks* 18(1): 99.

Burke, H. and Smith, C. 2010. 'Vestiges of colonialism: Manifestations of the culture/nature divide in Australian heritage management', in P. M. Messenger and G. S. Smith (eds), *Cultural heritage management: A global perspective*. Gainesville, FL: University Press of Florida, pp. 21–37.

Cleere, H. 1995. 'Cultural landscapes as world heritage', *Conservation and Management of Archaeological Sites* 1(1): 63–68.

Daley, B. and Griggs, P. 2006. 'Mining the reefs and cays: Coral, guano and rock phosphate extraction in the Great Barrier Reef, Australia, 1844–1940', *Environment and History* 12(4): 395–433.

—— 2008. '"Loved to death": Coral collecting in the Great Barrier Reef, Australia, 1770–1970', *Environment and History*, 14(1): 89–119.

Deane, F., Wilson, C., Rowlings, D., Webb, J., Mitchell, E., Hamman, E., ... Grace, P. 2018. 'Sugarcane farming and the Great Barrier Reef: The role of a principled approach to change', *Land Use Policy*, 78: 691–8.

Douvere, F., & Badman, T. (2012). *Report on the Reactive Monitoring Mission to the Great Barrier Reef (Australia), 6–14 March 2012*. Paris: UNESCO

Fabricius, K. E., Okaji, K. and De'ath, G. 2010. 'Three lines of evidence to link outbreaks of the crown-of-thorns seastar *Acanthaster planci* to the release of larval food limitation', *Coral Reefs* 29(3): 593–605.

Great Barrier Reef Foundation. 2017. *Resilient reefs successfully adapting to climate change*, https://www.barrierreef.org/uploads/Final%20report%20Resilient%20Reefs%20Successfully%20Adapting%20to%20Climate%20Change%20program.pdf.

Harrison, R. 2015. 'Beyond "natural" and "cultural" heritage: Toward an ontological politics of heritage in the age of Anthropocene', *Heritage & Society*, 8(1): 24–42.

Lee, E. 2016. 'Protected areas, country and value: The nature–culture tyranny of the IUCN's Protected Area Guidelines for Indigenous Australians', *Antipode*, 48(2): 355–74.

Lilley, I. and Pocock, C. 2018. 'Australia's problem with Aboriginal World Heritage', *The Conversation*, 13 December, https://theconversation.com/australias-problem-with-aboriginal-world-heritage-82912.

Lowenthal, D. 2005. 'Natural and cultural heritage', *International Journal of Heritage Studies* 11(1): 81–92.

McCalman, I. 2017. 'Linking the local and the global: What today's environmental humanities movement can learn from their predecessor's successful leadership of the 1965–1975 war to save the great barrier reef', *Humanities* 6(4): 77.

McIntyre-Tamwoy, S. 2004. '"My Barrier Reef": Exploring the Bowen community's attachment to the Great Barrier Reef', *Historic Environment* 17(3): 19–28.

Miller, I., Sweatman, H., Cheal, A., Emslie, M., Johns, K., Jonker, M. and Osborne, K. 2015. 'Origins and implications of a primary crown-of-thorns starfish outbreak in the Southern Great Barrier Reef', *Journal of Marine Biology*, doi: 10.1155/2015/809624.

Olsson, P., Folke, C. and Hughes, T. P. 2008. 'Navigating the transition to ecosystem-based management of the Great Barrier Reef, Australia', Proceedings of the National Academy of Sciences 105(28): 9489–94.

Pocock, C. 2002. 'Sense matters: Aesthetic values of the Great Barrier Reef', *International Journal of Heritage Studies* 8(4): 365–81.

—— 2004. 'Real to reel reef: Space, place and film at the Great Barrier Reef', in T. Ferrero-Regis and A. Moran (eds), *Placing the moving image* (3rd ed.). Brisbane: Griffith University, pp. 53–68.

—— 2006. 'Sensing place and consuming space: Changing visitor experiences of the Great Barrier Reef'. In K. Meethan, A. Anderson and S. Miles (eds), Tourism, consumption and representation: Narratives of place and self. Wallingford: CABI, pp. 94–112.

—— 2009. 'Entwined histories: Photography and tourism at the Great Barrier Reef'. In M. Robinson and D. Picard (eds), *The framed world: Tourism, tourists and photography*. Farnham: Ashgate, pp. 185–97.

—— 2010. 'A playground for science: Great Barrier Reef', Queensland Historical Atlas, 2009–10, http://www.qhatlas.com.au/content/playground-science-great-barrier-reef.

—— 2020. *Visitor encounters with the Great Barrier Reef: Aesthetics, heritage, and the senses*. London: Routledge.

—— 2022. 'Visualising heritage landscapes in future: aesthetics, embodiment, and meaning'. In C. Goetcheus & S. Brown (eds), *Routledge Handbook of Cultural Landscapes*, in press.

Rössler, M. 2002. 'Linking nature and culture: World Heritage cultural landscapes'. In P. Ceccarelli and M. Rössler (eds), *Cultural landscapes: The challenges of conservation*. *Paris*: UNESCO, pp. 10–15, https://unesdoc.unesco.org/ark:/48223/pf0000132988.

—— 2006. 'World Heritage cultural landscapes: A UNESCO flagship programme 1992–2006'. *Landscape Research* 31(4): 333–53.

Smith, A. 2013. 'People and their environments: Do cultural and natural values intersect in the cultural landscapes on the World Heritage List?'. In D. Frankel, S. Lawrence and J. Webb (eds), *Archaeology in environment and technology: Intersections and transformations*. London: Routledge, pp. 181–203.

Suggett, D. J., Camp, E. F., Edmondson, J., Boström-Einarsson, L., Ramler, V., Lohr, K. and Patterson, J. T. 2019. 'Optimizing return-on-effort for coral nursery and out-planting practices to aid restoration of the Great Barrier Reef', *Restoration Ecology* 27(3): 683–93.

Tweed, W. C. 2010. *Uncertain path: A search for the future of national parks*. Berkeley, CA: University of California Press.

UNESCO World Heritage Centre. 2019. *Operational Guidelines for the Implementation of the World Heritage Convention.* Paris: UNESCO.

Westcott, D. A., Fletcher, C. S., Kroon, F. J., Babcock, R. C., Plagányi, E. E., Pratchett, M. S. and Bonin, M. C. 2020. 'Relative efficacy of three approaches to mitigate crown-of-thorns starfish outbreaks on Australia's Great Barrier Reef', *Scientific Reports* 10(1): 1–12.

Wright, J. 1977. *The coral battleground.* Sydney: Angus & Robertson.

Celmara Pocock is the Director of the Centre for Heritage and Culture and Associate Professor of Anthropology and Heritage Studies at the University of Southern Queensland. She is a leading heritage scholar with interests in social significance and community heritage, including Indigenous heritage, aesthetics and senses of place, storytelling and emotion, and the intersections between heritage and tourism. Her monograph *Visitor Encounters with the Great Barrier Reef: Aesthetics, Heritage, and the Senses* was published by Routledge in 2020.

Sounds of silence

Diane Tarte
ditarte@ozemail.com.au

It was the early 1980s on a warm summer's evening on North West Island, located in the Capricornia Bunker Group towards the southern end of the Great Barrier Reef. I had some time to myself and was wandering along the beach at sunset. Looking up, I realised there were thousands and thousands, if not hundreds of thousands, of birds — wedge-tailed shearwaters, in fact — circling the island as they returned to their underground nests and their mates and chicks after a day of feeding and cruising the air currents. What was so special about this? After all, it happens every summer's evening on many Reef sand cays. It was special for me because I suddenly realised that this huge sweep of birds was flying past in total silence ... the only sound was an occasional wing dipping into the sea.

Anyone who has spent any time in summer on the Reef's southern vegetated sand cays will regale you with stories of the moaning shearwaters ... it goes on all night and can be the making of nightmares. Yet here they were in their teeming thousands in total silence.

It's not often that silence is a fond memory of the Reef. Generally, there's always something happening, either above or below water. More often, silence is the harbinger of a problem — as happened when I went snorkelling on reefs off Cairns after the 2016 mass coral bleaching event. Certainly there were some patches of pretty coral reef left with the usual array of darting fish and invertebrates, but it was the silent swathes of collapsing dead corals bereft of their usual cluster of co-inhabitants that filled my memories of that trip.

For someone who has spent most of her life trying to ensure that the Reef has a future, and that this future is as good as its past, the last five years have been hard. Just when it seemed that modest progress was being made with controlling pollution from the catchments, and on-water management was generally coping with most of the *in situ* challenges apart from the depredations of the crown-of-thorns starfish (COTS), the Reef suddenly experienced the bullets it had been dodging — three marine heatwaves triggering extensive mass coral bleaching events in just five years.

While the computer modelling by coral reef and climate experts showed it was inevitable that sometime soon the Reef would experience extensive mass coral bleaching events, before 2016 these had occurred at intervals that meant there was some reasonable time to recover. Instead, now it was three events in five years, whereas the recovery time for coral reefs is ten to fifteen years between events. Every summer now, I watch the weather patterns in January, February and March and heave a huge sigh of relief when the bullet passes again.

The solution is obvious, and has been for the past fifty years: reduce our greenhouse gas emissions, both nationally and globally. Yet Australia has been locked into ridiculous political climate wars, while sensible public policy measures such as a tax on carbon and incentivising industries and communities to shift to renewable and sustainable energy alternatives have been ridiculed and even demonised. However, many Australians are showing our government what is needed, and international mechanisms such as carbon tariffs on imports and an 'in danger' World Heritage Convention listing for the Reef may force the long overdue U-turn on national climate policy.

We should also take pride in the measures we have taken to protect the Reef over the past fifty years. Because we had the foresight in 1975 to declare the largest marine protected area at that time, and to increase the levels of protection and management over the past decades, some areas of the Great Barrier Reef are inherently more resilient than coral reefs in many other countries. We also understand better the when, where and what to do for continuing to reduce the impacts of catchment sourced pollution. Moreover, science, advanced electronics and artificial intelligence are showing some promise in suppressing outbreaks of COTS and limiting some of the impacts of rising sea temperatures on coral reef communities.

The 1969 slogan 'Save the Barrier Reef' is still as relevant today as it was then. The original 'Save the Barrier Reef' campaign emerged due to mining applications for limestone and then oil and gas extraction from the Reef. The future of the Reef is still dependent on limiting the footprint of the oil and gas industry, but not just locally now — this time it is globally.

I take pride in what has been achieved for the Reef; I despair for what we have lost, when the risks were well known; and I have a glimmer of hope that there is a small chance of providing the breathing space for the Reef to sustain itself and recover some of its lost vigour.

Diane Tarte is Director of Marine Ecosystem Policy Advisors, providing advice on policy and programs addressing research and management of marine, coastal and catchment areas. Since the late 1970s, Diane has been involved in advocating for protection of the Great Barrier Reef and has undertaken field inventory work on Reef islands and cays, and Queensland coastal tidal wetland systems.

Coal versus coral: Australian climate change politics sees the Great Barrier Reef in court

Claire Konkes, Cynthia Nixon, Libby Lester and Kathleen Williams

Claire.Konkes@utas.edu.au
Cynthia.Nixon@utas.edu.au
Elizabeth.Lester@utas.edu.au *and*
Kathleen.Williams@utas.edu.au

Abstract

The likelihood that climate change may destroy the Great Barrier Reef has been a central motif in Australia's climate change politics for more than a decade as political ideologies and corporate and environmental activism draw or refute connections between the coal industry and climate change. The media fuel this debate because in this contest, as ever, the news media always do more than simply report the news. Given that the Reef has also been central to the evolution of Australia's environmental laws since the 1960s, it is not surprising that the Reef is now a leading actor in efforts to test the capacity of our environmental laws to support action on climate change. In this contribution, we examine the news coverage of the Australian Conservation Foundation's (ACF) 2015 challenge to Adani's Carmichael coal mine to observe the discursive struggle between the supporters and opponents of the mine. Our analysis of the case shows that while the courts are arenas of material and symbolic contest in the politics of climate change in Australia, public interest environmental litigants struggle both inside and outside the courts to challenge the privileging of mining interests over the public interest.

The Great Barrier Reef (the Reef) has been central to the evolution of Australian environmental policy for more than half a century (Bowen and Bowen 2002; Foxwell-Norton and Konkes 2018; Foxwell-Norton and Lester 2017). More recently, the threat of climate change to the Reef has become a key motif in Australia's climate change politics (Hobbs and McKnight 2014; Konkes and Foxwell-Norton 2021; Nixon et al. 2021), which, more than in many other countries, is strongly influenced by political ideologies and corporate and environmental activism, flamed by media always doing more than simply reporting the news (Lester 2019). Although 'climate change is fundamentally a political issue' (Carvalho, van Wessel and Maeseele, 2017: 124), environmental non-government organisations (eNGOs) are increasingly turning to the court system to challenge mining and other practices that contribute to carbon emissions (Konkes 2018;

Schuijers and Young 2021). At the current juncture, where global and local media amplify the spectacle of the Reef and its politics, we examine the intersections of news media, activism, politics and environmental law to better understand the communication strategies enlisted when climate change goes to court and the representation of such strategies in news media.

Australia has the third-largest reserves of coal in the world and is the second largest coal exporter (Geoscience Australia 2021). In 2019–20, coal exports were worth A$55.2 billion and the industry directly employed more than 50,000 workers, mainly in Queensland (Minerals Council of Australia 2021). The coal mining industry is therefore a major stakeholder of the Australian and state governments (Bacon and Nash 2012; Lester 2017; Readfearn 2015; Stutzer et al, 2021). However, the future of the coal industry lies in the speed of the global transition to renewable energy sources, and the pace of global economic growth (Cunningham et al. 2019). Within this global context, developing the coal reserves in Central Queensland's Galilee Basin is variously represented as being an economic necessity or a retrograde step along the road to Australia's low-carbon future (Jolley and Rickards 2020). Of all the proposed projects or those already underway, Adani's Carmichael coal mine has received the most prominent opposition. As well as emphasising the contribution of coal to carbon emissions, opponents are critical of the viability of — and the public interest in — a project that requires government subsidies and tax concessions estimated to be A$4.4 billion (Smee 2019).

Opposition to the Carmichael coal mine, then, can be seen as an expression of despair and hope: despair for the apparent failure of governments to act on climate change and hope for a fossil-free future. At a local level, coal mining communities are caught between the familiar challenge of supporting economic benefit amid the competing voices in the decision-making processes (Lester 2019). These voices include state and federal governments, news media actors, the resource sector and a wide range of community groups, all of whom contain people who variously support or oppose the mine for a multitude of reasons. In these local Queensland communities and beyond, the Carmichael coal mine has come to symbolise the short-term economic (and political) benefits of coal mining and the longer-term advantages of a low-carbon future. As such, the mine has become enmeshed in Australia's climate change politics. Operating in this complex communication environment, those seeking action on climate change draw on heuristic tools, such as metaphor and symbol, to convey the abstract and arcane to the public (Hulme 2012). Here, the enduring visual spectacle and symbolic power of the Reef works to connect the act of mining and burning coal to the deleterious impact of climate change on natural wonders, such as the Reef. A coordinated international campaign to end Australia's coal industry (Hepburn et al. 2011, p. 6) identified the Reef as a one area of focus in public interest environmental litigation.

Our contribution examines how the Australian news media represent public-interest environmental litigation targeting action on climate change through an analysis of the ACF case initiated in 2015[1] — where the ACF argued that the minister should have considered the climate change impacts on the Great Barrier Reef from burning the mine's coal.

We first introduce the Carmichael coal mine as the focus of campaigns for climate change action and describe how the Reef has long been a symbol in Australian environmental campaigns. We then discuss the increasing use of litigation in climate change politics and draw on the ACF's 2015 case to analyse the news coverage by seven Australian news outlets to examine how news represented the discursive struggles to not only stop the mine, but also reform Australian environmental law. When news reporting of environmental conflict is observed as the result of the interlocking activities, strategies and communication between journalists and their sources in activism, industry and formal politics (Hutchins and Lester 2015), we can observe how the courts are not separate, but part of, the politics of climate change in Australia and elsewhere.

The Reef, Australia's climate wars and the law

The contradiction between Australia's international responsibility to protect the Reef and its political appetite for carbon-intensive industries is epitomised in the development of mining in Central Queensland's Galilee Basin. Exploration licences in the Basin have long been held by companies such as Clive Palmer's Waratah Coal and GVK Hancock Coal (Burt 2019), but the likelihood of mining in the Galilee Basin became a reality in 2010 when Adani Australia announced plans to build its Carmichael coal mine. Adani Australia, a subsidiary of the Adani Group, founded and chaired by Indian tycoon Gautam Adani, operated the Carmichael coal mine and the associated Carmichael Rail Project until changing its name to Bravus in 2020 (Zhou 2020). The project includes plans for open-cut and underground mines, large waste dumps and tailing dams across an estimated 44,000 hectares. It also requires considerable infrastructure, including 189 kilometres of rail to transport coal from the Carmichael coal mine to an export facility at Abbot Point Port, located on the shores of the Reef where coal will be shipped to India for use in Adani-owned coal-fired power stations (Lester 2017). Plans for A$900 million in federal funding for the rail that links the mine to the port were shelved in 2020 (Karp 2020). To date, the company has received A$3.2 million in federal funding, leaving the Carmichael coal mine and railway largely funded by Adani (Slezak 2020), but dependent on government subsidies and tax concessions (Smee 2019).

Since the 2010 announcement of the mine — and irrespective of the name change — 'Adani' has come to symbolise the large corporate ventures deliberately exploiting political processes to the detriment of Australia's public interest (Bradley 2019) and the 'systemic web of access and influence' binding the coal industry to Australian politics (Readfearn 2015). This web of interests is, of course, bigger than Adani. For example, Gina Rinehart, the principal of GVK Hancock Coal, is a board member of one of Australia's major news organisations, Nine Fairfax (Burt 2019), and a major donor to Australia's influential conservative thinktank, the Institute of Public Affairs (Seccombe 2018). Similarly, in the 2019 federal election, Clive Palmer's United Australia Party broke Australian election spending records on an advertising campaign widely credited with returning the incumbent conservative government to office (McCutcheon 2020). As Jolley and Rickards (2020: 18) note, 'coal is symbolic not just of old technology but of an old style of politics, in which the narrow interests of the rich and powerful trump those of the wider social good'. Campaigns that have confronted this 'old style of politics' have long known the power of the Reef to not only capture people's attention, but to compel support for action. The proclamation of the Reef as a World Heritage site in 1981 was a major win for Australia's emergent conservation movement, not least because the campaign to save the Reef from gas and petroleum exploration exposed the shortfall in relying on an ad hoc assemblage of local, state and Commonwealth laws to protect our natural places (Bowen and Bowen 2002; Konkes 2018).

From the success of early campaigns to save the Reef, Australia began formalising its environmental laws, notably the passing of the *Environmental Assessment (Impact of Proposals) Act* in 1974 and the *Environmental Protection and Biodiversity Conservation Act* (EPBC Act) in 1999. These laws marked the beginning of some cohesion in Australian environmental policy and public interest law (Bowen and Bowen 2002), and the reification of an ecological perspective into its political and legal decision-making processes (Wickham and Goodie 2013). That said, climate change did not have the political saliency it does today. With a few exceptions — notably the 2017 decision on the Rocky Hill Mine Project (see McGrath 2021) — the EPBC Act has proven to be despairingly inadequate for addressing the complex and cumulative impacts of climate change (Grech et al. 2016). This absence was identified in the first review of the Act (Commonwealth of Australia 2009: 146), which noted the need for 'better systems' to address 'the additional pressure of climate change'. The most recent review (Samuel 2020) was more damning: Australia's key environmental laws were 'outdated' and efforts to address climate change described as 'ad hoc, rather than a key national priority' (Samuel 2020: 17).

With systemic action to address climate change requiring effective legal frameworks (Goodie and Wickham 2002), it is little wonder that those campaigning for action on climate change are turning their attention to the courts in Australia and elsewhere (Preston 2016). This trend sees litigation as a nascent expression of a struggle for power as much as law (Murphy and McGee 2018, p.166). In this context, litigation is being undertaken in Australia to not only stop projects such as the Carmichael coal mine, but to fundamentally reform a legal framework that treats the natural world and its systems as property to be owned and exploited and, further, those acting in the public interest to protect the environment being treated as 'criminals who infringe upon the property rights of others' (Cullinan 2008). That the legal system considers each case on its individual merits is another challenge because it enables the denial of responsibility for cumulative impacts to be a defence against challenges to new projects and makes arguing the case for the cumulative effects of climate change difficult (Lowe 2004; McGrath 2008). As such, climate change action requires a necessary shift 'from anthropocentric to ecocentric' legal frameworks (Barcan 2019). While the finer adjustments to policy and law are of interest to a few, the importance of action on climate change is understood by many. For this reason, the power of the symbolic, such as a wondrous reef, continues to be used to engage public attention and mobilise political demand for action that requires, when the campaign caravan moves on, the less-spectacular changes to governance frameworks. In this process, the Reef works as both a literal location and a heuristic instrument in the discursive struggle for a legal framework able to respond to the challenge of climate change.

Mediatising climate change law

The increasing shift towards public interest environmental litigation and climate change litigation is opening a new arena from which to observe how power is enacted in Australia's climate change politics (Preston 2016). Such litigation can be seen through the lens of Brett Hutchins and Libby Lester's (2015) model of 'mediatised environmental conflict', which theorises the interlocking networks and complex interactions between activist strategies, news reporting, industry activities and formal politics and decision-making processes. In these contests, those who can exercise control of what is visible (or invisible) have a strategic advantage. The model builds on Simon Cottle's (2006: 9) work on 'mediatised conflict', which describes the role of media as actively 'doing something' over and above disseminating ideas, images and information. Notably, the term 'mediatization' draws attention to media as an independent social institution with its own sets of rules to which others adapt their practices (Hjarvard 2013) as well as the resulting processes in the communicative construction of our cultural and social reality (Couldry and Hepp 2013).

The processes of mediatised environmental conflict help us to observe how public-interest environmental litigation leads to law being 'shaped and transformed by actors and agencies outside of those sites that we normally associate with the formation of law' (Goodie and Wickham 2002: 37). For instance, even when unsuccessful, litigation can be used to test the public mood for legal reform (Schuijers and Young 2020) or signal to stakeholders the existence and extent of public concern about an activity. In this context, the strategic use of litigation is a communicative action, and thus of interest to interdisciplinary researchers interested in the processes by which the enactment of power and those who challenge it are communicated. That said, public interest environmental litigation is almost the antithesis to the spectacle of mediatised environmental conflict because courts seldom engage directly with news media: they rarely issue press releases or hold media conferences, nor do they conduct interviews about case decisions. Instead, the primary form of communication is written documents using legal language for legal audiences, and journalists are left with the task of translating events to the wider public. The disinclination of those in the legal system to engage with news media, and criticisms of the inaccuracy in news reporting, contribute to a culture of antagonism between lawyers and journalists (Breit 2011). However, like other forms of news production, court reporting draws on sources to comment and interpret the day's events, demanding that both journalistic practice *and* the communication strategies of sources be considered when examining the translation of legal issues in news (Konkes and Lester 2016). When news reporting of environmental litigation is observed as the result of the interlocking activities and communication strategies of journalists and their sources, we can observe the courts as an arena in the politics of climate change, where even an unsuccessful case can serve wider purposes.

Taking coal to court

The campaign to stop coal mining in Australia first came to prominence when the strategy, *Stopping the Australian Coal Boom* (Hepburn et al. 2011) was leaked to news media. Supported by a consortium of international environmental groups and their financial backers, it included an explicit push for 'climate change law in Queensland and New South Wales so future climate change cases are more likely to succeed' (Hepburn et al. 2011, p. 6) and signalled to industry and government a major resistance to new coal developments in Australia. Nine legal cases using environmental law to stop the Carmichael coal mine were launched between 2014 and 2019, with further cases using native title law (e.g. see Wangan and Jagalingou Family Council 2019). Much has been written about these environmental law cases, including their coverage by news media (e.g. Jolley and Rickards 2020; Konkes 2018; Nixon et al. 2021).

We contribute to this growing body of work by examining the 2015 judicial review initiated by the Australian Conservation Foundation (the ACF case).[2] Under the EPBC Act, a judicial review considers whether an exercise of statutory power by government was exercised lawfully. In this case, the ACF alleged the Minister had failed to properly take into account that carbon emissions created by Australian coal burned in India was inconsistent with Australia's obligation to protect the Reef. Although the ACF lost the case, it was noteworthy within the legal sphere for the detail in which the court was asked to consider evidence of climate change (Murphy and McGee 2018) and it highlighted the limitations of Australian environmental law and policy to challenge the merit of large projects (McGrath 2021). Of interest to us was the efficacy of such litigation as a strategy to not only stop a mine, but also communicate the need for law reform.

Seven Australian news media outlets ranging across the 'political spectrum' — *The Australian*, *ABC News*, the *Australian Financial Review*, *Guardian Australia*, the *Courier-Mail*, the *Sydney Morning Herald* and the *Central Queensland News* — were selected for content analysis, including identifying and counting the news sources quoted in texts (Tankard 2003). From a corpus of 46 news stories covering the case, we drew on Kitzinger's (2009) emphasis on 'language devices' to examine how, when and by whom frames are used. Using critical discourse analysis (Carvalho 2008), we identified five frames used to describe the case: (1) court conflict; (2) activist tactic; (3) public right; (4) bureaucracy; and (5) criminality. A semi-structured interview with Australian Conservation Foundation's then senior media adviser, Josh Meadows, enabled us to compare the ACF's communication strategy with the resulting news coverage. Supplementing observations with interviews is an orthodox approach in source-media research that enables researchers to assess the success and failure of sources to influence agendas (Anderson 1997; Lester and Hutchins 2012).

In the public interest or just a folly?

There were several elements to ACF's media strategy (Meadows, interview 2018), which can be seen at work in news coverage. The ACF flew journalists over the Reef to observe first-hand the effects of the 2016 mass bleaching event in the month before the case was heard (Meadows, interview 2018), which may have informed how some journalists connected the devastation of the Reef to the burning of coal (e.g. Slezak 2016a). In its communication with journalists, the ACF emphasised the 'novelty' of the case as a news value, and some reporters reported how the 'historic, landmark' case was a test to strengthen environmental laws (e.g. Robertson 2015; van Vonderen 2015). Similarly, the ACF argued that the mine was illegal, or at least a crime against nature. Only one report drew on this frame. *Guardian Australia* (e.g. Robertson 2015) used the term 'illegal' to describe the mine; in other news reporting on the Adani conflict, terms associated with criminality, such as 'trespass', were used to describe the actions of activists, rather than an administrative 'error in law' involving the Minister not taking into consideration the impact of climate change on the Reef. The ACF also drew on its association with successful campaigns by comparing this case with Tasmania's Franklin River and the World Heritage listing of the Reef (e.g. Australian Associated Press 2015; Robertson 2015). Anticipating that mine supporters — including news media — would deploy strategies to 'paint this as greenies trying to delay and obstruct' (Meadows, interview, 2018), the ACF emphasised how it wanted to 'stop the mine', not just 'disrupt and delay' it. Despite these efforts, our findings show that media coverage largely framed the litigation in terms of being an *activist tactic*, rather than an attempt to stop a large project and reform Australia's much-criticised environmental laws.

The dominant *activist tactic* framing can partly be explained by the coverage of the case by *The Australian* and the *Courier-Mail*, both owned by Murdoch's News Corp. When the ACF lodged papers in the Federal Court in Brisbane in November 2015, *The Australian* and the *Courier-Mail* connected the case to the *Stopping the Australian Coal Boom* strategy and described it as an instance of 'guerrilla lawfare' (McKenna and Maher 2015). An editorial in the *Courier-Mail* (2015) said 'enough is enough' and an editorial in *The Australian* (Johns 2015) raised questions about the funding sources for the legal action and accused the ACF of 'purposely set[ting] out to harm other Australians'. Other outlets drawing on industry sources also contributed to this frame. For instance, *Guardian Australia* (Robertson 2015) included comments from an Adani spokesman about the 'endless' (used three times) nature of litigation. While the connection between coal, climate change and the Reef was not denied, the argument for Australia to take responsibility for emissions was trivialised by some sources, including the Queensland Resources Council chief executive officer Michael Roche — reported by ABC News — suggesting that to ask Australia to take responsibility for Adani's emissions was akin to asking Saudi Arabia to take responsibility for fuel combusted in Australian cars (van Vonderen 2015).

The judicial hearing was heard across two days before the Federal Court in May 2016. While two organisations reported both days, others did not report at all. A sense of *court conflict* dominated the headlines. Some reporters detailed the climate change evidence given before the court, and extended this connection by linking the case to recent reports of coral bleaching on the Reef (Slezak 2016b) or quoting an ACF spokesperson outside court (Frost 2016).

Media attention returned when the Federal Court dismissed the case on 29 August 2016. Adani sources drew directly on the discourses of the *activist tactic* and *bureaucracy* frames in their response to the dismissal. Notably, *The Australian* (Schliebs 2016) quoted the Queensland Resources Council chief executive Ian Macfarlane (a former resources minister for two Liberal governments) criticising the ACF for using 'activists' tactics [that] mean that the only jobs being created are for lawyers'. Others reported ACF claims that Australia's environmental laws were 'broken', 'weak' and insufficient, and the ACF's calls for the federal Environment Minister to revoke the approval (Robertson 2015). Several actors sought to maintain momentum for change: the ACF organised a public protest, while the mine's supporters, such as federal MP George Christensen (2016), called for closing the 'legal loopholes being exploited by "green groups"'.

The predominance of news stories framing the ACF's legal challenge as *an activist tactic* was not surprising: the ideological cultures and stances of conservative outlets in Australia, and elsewhere, have contributed to their silence on climate change (McAllister et al. 2021). Notably, Rupert Murdoch's media empire — particularly his Australian newspaper *The Australian* — have for decades maintained a strategy to advance Murdoch's political views in which climate-change denial looms large (Bacon and Nash 2012; McKnight 2005, 2012; McKnight and Hobbs 2017). However, other outlets, such as *Guardian Australia*, also contributed to this framing by quoting the opinions and more colourful phrasing of industry sources. While providing a mix of views, or 'balance' is a mainstay of objective reporting, the choice of phrasing can also serve to amplify informational bias, described by some as the easy pitfall of 'false balance' (Fahy 2017).

Reporting of the case judgement provides a stark contrast — and insight into the ideology — between the various news organisations. The 52-page written case judgement[3] of the case mentions the 'Great Barrier Reef' 78 times, 'climate change' 34 times and 'greenhouse gas' 32 times, which gives some indication of the significance to climate change and environmental protection attached to the case. However, news media translation of the case judgment, through the choice of words and absence of others, either enhanced the connection between the mine and the Reef or removed it entirely. The decision to include or ignore this connection influenced how the case was framed as a matter of *public interest* or as an *activist tactic*. For instance, despite the legal arguments about Australia's international obligations to protect the Reef from climate change, *The Australian* (Schliebs 2016) avoided mentioning the 'Great Barrier Reef', 'climate change', 'global warming', or 'greenhouse gas emissions' entirely in its coverage, and instead described the case as challenging the 'impact to the environment from burning coal'. In comparison, the *Sydney Morning Herald* (Hannam 2016) drew a connection between climate change and the recent coral bleaching of the Reef, including scientific evidence and video footage taken when the ACF took journalists to see the 2016 bleaching event.

The dismissal of the case and the reporting of the judge's decision were followed by news coverage of the order to pay costs on 9 September 2016. The *Courier-Mail* (McCarthy 2016) covered the judge's order for the ACF to pay 40 per cent of Adani's costs and 70 per cent of the government's costs, and quoted Queensland Resources Council chief executive Michael Roche describing the action as 'folly' that would be difficult to justify to its members. In contrast, the judge's order to limit the amount the ACF had to pay was noted by legal professionals, such as McKinnon (2016), for being a 'rare and significant departure from the usual' because it acknowledged the litigant's 'credibility [and] representation of the Australian public and the public interest'. As McGrath (2021) observed, while 'procedural issues like costs seem boring', costs are a critical issue for public interest litigants not pursuing commercial interests. News coverage returned two weeks later when the ACF announced it would appeal the decision, calling the mine's approval 'a licence to kill' and comparing building a coal mine to murder (Robertson 2016). News of appeal was countered by the story that Adani was scaling back its Carmichael coal mine investment, with the *Financial Review* attributing the decision to the 'anti-coal lobby's hydra-headed campaign', rather than financial or legal viability of the mine (Stevens 2016).

Hope and despair: Defending our future in court

From the outset, the ACF (2016) said the case would be difficult to win, admitting that the purpose of the litigation was as much about highlighting Australia's 'failed environmental laws' and 'failure to act on climate change' as it was about stopping the mine. When the application for judicial review was dismissed because the court found the Minister made no error of law in approving the mine, ACF and others (McGrath 2021) noted the irony: although the Australian government acknowledged climate change threatened the Reef and that the Carmichael coal mine would produce significant carbon emissions, the country's environmental laws could not be applied to stop the mine. Despite being unsuccessful, the case was significant for the detailed way in which the court was asked to consider evidence of climate change (Murphy and McGee 2018) and its exposure to how scientific evidence was not effectively and transparently incorporated into the approvals process (Currell et al. 2020).

A decade of diverse and persistent campaigning against the Carmichael coal mine has made Adani synonymous with Australia's inaction on climate change, caused in large part by political decision-making processes that privilege the interests of the fossil fuel industry over the public interest. This imbalance of power is an 'old style of politics' continually being reconstituted to adapt to a changing global context. Efforts to challenge old ways of doing politics require communicative acts that not only imagine a new way of doing politics, but that articulate the steps required to get there. Efforts to ensure our policies and laws incorporate scientific evidence and other ecocentric perspectives are small but important steps. How legal arguments are represented in media provides the opportunity for them to be tested in the court of public opinion, as well as inside court.

In this article, we have shown that news coverage of such cases reflects the communication strategies of all parties, with the choices of sources and quotes by journalists contributing to framing climate change litigation, in the case described here, as an act of activism, rather than being undertaken in the public interest. Despite the symbolic power of the Reef to represent what we stand to lose if we do not act on climate change, such framing risks diminishing the public interest in action on climate change. As ever, the symbolic power of the Reef is being used to direct us to imagine what we stand to lose if we fail to act to address climate change. Challenging the old way of doing politics, including legal orthodoxies that impede action on climate change, is a major undertaking — not to mention the added challenges our cultural imaginations also present in which there are many ways to interpret the symbolic meaning of the Reef. Despite the complexity of the communicating climate change action, news framing that reduces such efforts as the antics of activists does little other than serve those profiting from the old way of doing business.

Acknowledgements

This research was in part funded by an Australian Research Council Project DP150103454 'Transnational Environmental Campaigns in the Australia-Asian Region' and supported by an Australian Government Research Training Program (RTP) scholarship.

Notes

1 *Australian Conservation Foundation Incorporated v Minister for the Environment* [2016] FCA 1042 and 1095.
2 *Australian Conservation Foundation Incorporated v Minister for the Environment (No. 2)* [2016] FCA 1095.
3 *Australian Conservation Foundation Incorporated v Minister for the Environment* [2016] FCA 1042 and 1095.

References

Australian Associated Press. 2015. 'Adani mine faces another challenge', *The Weekend Australian*, 10 November, https://www.theaustralian.com.au/business/business-spectator/adani-mine-faces-another-challenge/news-story/3af7a4986539b3c5b56fc8a91d552226.

Anderson, A. 1997. *Media, culture and the environment*. London: Routledge.

Australian Conservation Foundation. 2016, 'ACF versus the Environment Minister – our Adani court case explained', https://www.acf.org.au/acf_vs_environment_minister.

Bacon, W. and Nash, C. 2012, 'Playing the media game', *Journalism Studies* 13(2): 243–58.

Barcan, R. 2020. 'The campaign for legal personhood for the Great Barrier Reef: Finding political and pedagogical value in a spectacular failure of care', *Environment and Planning E: Nature and Space* 3(3): 810–32.

Bowen, J. and Bowen, M. 2002. *The Great Barrier Reef: History, science, heritage*. Melbourne: Cambridge University Press.

Bradley, J. 2019. 'How Australia's coal madness led to Adani', *The Monthly*, April, https://www.themonthly.com.au/issue/2019/april/1554037200/james-bradley/how-australia-s-coal-madness-led-adani?cb=1612648184

Breit, R. 2011. *Professional communication: Legal and ethical issues* (2nd ed.). London: LexisNexis Butterworths.

Burt, J. 2019. 'Adani could be "ice-breaker" for six more proposed Galilee Basin mines, resources body says', *ABC News*, 12 June, https://www.abc.net.au/news/2019-06-12/adani-approval-could-be-galilee-basin-ice-breaker/11194510

Carvalho, A. 2008. 'Media(ted) discourse and society: Rethinking the framework of critical discourse analysis', *Journalism Studies* 9: 161–77.

Carvalho, A., Van Wessel, M. and Maeseele, P. 2017. 'Communication practices and political engagement with climate change: A research agenda', *Environmental Communication* 11(1): 122–35.

Christensen, G. 2016, 'MP urges Labor to support job creating Bill', media release, 8 February, https://www.georgechristensen.com.au/bygeorge/mp-urges-labor-to-support-job-creating-bill.

Commonwealth of Australia. 2009. *First Independent Review of the Environment Protection and Biodiversity Conservation Act 1999: Interim report*. Canberra: Australian Government, https://www.environment.gov.au/system/files/resources/5d70283b-3777-442e-b395-b0a22ba1b273/files/interim-report.pdf.

Cottle, S. 2006, *Mediatized conflict: Developments in media and conflict studies*. London: McGraw-Hill.

Couldry, N. and Hepp, A. 2013. 'Conceptualizing mediatization: Contexts, traditions, arguments'. *Communication Theory* 23: 191–202.

Courier-Mail. 2015. 'Balancing the rights of developers and climate',10 November, p. 22.

Cullinan, C. 2008. 'If nature had rights', *Orion Magazine*, https://orionmagazine.org/arti cle/if-nature-had-rights.

Cunningham, M., Van Uffelen, L. and Chambers, M. 2019. 'The changing global market for Australian coal', *Reserve Bank of Australia Bulletin*, September, pp. 28–38, https://www.rba.gov.au/publications/bulletin/2019/sep/the-changing-global-market-for-australian-coal.html.

Currell, M. J., Irvine, D. J., Werner, A. D. and McGrath, C. 2020. 'Science sidelined in approval of Australia's largest coal mine', *Nature Sustainability* 3(8): 644–9.

Fahy, D. 2017. 'Objectivity as trained judgment: How environmental reporters pioneered journalism for a "post-truth" era'. *Environmental Communication*, 12(7): 855–61.

Foxwell-Norton, K. and Konkes, C. 2018. 'The Great Barrier Reef: News media, policy and the politics of protection', *International Communication Gazette*, 81(3): 211–34.

Foxwell-Norton, K. and Lester, L. 2017, 'Saving the Great Barrier Reef from disaster, media then and now', *Media Culture and Society* 39: 568–81.

Frost, P. 2016. 'Lawyers: Carmichael emissions 3x more than reduction target', *Central Queensland News*, 3 May, https://www.cqnews.com.au/news/Lawyers-Carmichael-emissions-3x-more-than-reducti/3015419.

Geoscience Australia 2021. 'Coal', https://www.ga.gov.au/digital-publication/aecr2021/coal.

Goodie, J. and Wickham, G. 2002. 'Calculating "public interest": Common law and the legal governance of the environment', *Social & Legal Studies* 11(1): 37–60.

Grech, A., Pressey, R. L. and Day, J. C. 2016. 'Coal, cumulative impacts, and the Great Barrier Reef', *Conservation Letters* 9(3): 200–7.

Hannam, P. 2016. 'Australian Conservation Foundation loses Federal Court case on Adani coal', *Sydney Morning Herald*, 29 August, https://www.smh.com.au/environment/climate-change/australian-conservation-foundationloses-federal-court-case-on-adani-coal-20160829-gr3au2.html.

Hjarvard, S. 2013. *The mediatization of culture and society*, London: Routledge.

Hepburn, J., Burton, B. and Hardy, S. 2011. 'Stopping the Australian coal export boom: Funding proposal for the Australian anti-coal movement', *Media Watch*, ABC TV, https://www.abc.net.au/mediawatch/transcripts/1206_greenpeace.pdf.

Hobbs, M. and McKnight, D. 2014. '"Kick this mob out": The Murdoch media and the Australian Labor Government (2007 to 2013)', *Global Media Journal*, 8(2): 1–13.

Hulme, M. 2012. 'How climate models gain and exercise authority'. In K. H. Hastrup and M. Skrydstrup (eds), *The social life of climate change models*. London: Routledge, pp. 40–54.

Hutchins, B. and Lester, L. 2015. 'Theorizing the enactment of mediatized environmental conflict', *International Communication Gazette* 77(4): 337–58.

Johns, G. 2015. 'Making people's lives tougher is hardly an act of charity', *The Australian*, 9 November, p. 12.

Jolley, C. and Rickards, L. 2020. 'Contesting coal and climate change using scale: Emergent topologies in the Adani mine controversy', *Geographical Research*, 58(1): 6–23.

Karp, P. 2020. 'Adani coalmine won't get federal rail funding, Liberal minister says', *The Guardian*, 4 February, https://www.theguardian.com/environment/2018/feb/04/adani-coalmine-wont-get-federal-rail-funding-liberal-minister-says.

Kitzinger, J. 2009. 'Framing and frame analysis'. In E. Devereux (ed.), *Media studies: Key issues and debates*. London: Sage, pp. 134–61.

Konkes, C. 2018, 'Green lawfare: Environmental public interest litigation and mediatized environmental conflict', *Environmental Communication*, 12(2): 191–203.

Konkes, C. and Foxwell-Norton, K. 2021. Science communication and mediatised environmental conflict: A cautionary tale. *Public Understanding of Science*, 30(4): 470–83.

Konkes, C. and Lester, L. 2016. 'Justice, politics and the social usefulness of news', *Crime, Media, Culture* 12(1): 17–35.

Lester, L. 2017. 'Transnational discourses of risk and responsibility from Australia: Coal, pollution and the Great Barrier Reef'. In G. D. Hook, L., Lester, M Ji, K. Edney and C. G. Pope (eds), *Environmental pollution and the media: Political discourses of risk and responsibility in Australia, China and Japan*. London: Routledge, pp. 45–74.

—— 2019. *Global trade and mediatised environmental protest: The view from here*. London: Palgrave Macmillan.

Lester, L. and Hutchins, B. 2012. 'The power of the unseen: Environmental conflict, the media and invisibility', *Media, Culture & Society*, 34(7): 847–63.

Lowe, I. 2014. 'Wild law embodies values for a sustainable future'. In M. Maloney and P. Burdon (eds), *Wild law in practice*. London: Routledge, pp. 3–16.

McAllister, L., Daly, M., Chandler, P., McNatt, M., Benham, A. and Boykoff, M. 2021. 'Balance as bias, resolute on the retreat? Updates & analyses of newspaper coverage in the United States, United Kingdom, New Zealand, Australia and Canada over the past 15 years', *Environmental Research Letters* 16(9), doi: 10.1088/1748-9326/ac14eb.

McCarthy, J. 2016. 'Australian Conservation Foundation ordered to pay costs for Adani case', *Courier-Mail*, 9 September, https://www.couriermail.com.au/news/queensland/crime-and-justice/australian-conservation-foundation-ordered-to-pay-costs-for-adani-case/news-story/20a030d675a2edb873a9888fd1e152ea.

McCutcheon P. 2020, 'How is Clive Palmer maintaining his political influence at the Queensland election?', *ABC News*, 28 October, https://www.abc.net.au/news/2020-10-28/qld-election-how-clive-palmer-maintaining-political-influence/12814534.

McGrath, C. 2021. 'Carmichael Coal ('Adani') mine cases in the Federal Court', *Environmental Law* website, http://envlaw.com.au/carmichael-coal-mine-federal-court.

McKenna, M. and Maher, S. 2015. 'Risk to reef new front in coal campaign', *The Australian*, https://www.theaustralian.com.au/nationalaffairs/climate/risk-to-great-barrier-reef-new-front-in-adani-mine-coal-campaign/newsstory/64ddff50142413e1301dad1b2124d087.

McKinnon, E. 2016. '*ACF versus the Environment Minister – our Adani court case explained*'. Australian Conservation Foundation, https://www.acf.org.au/acf_vs_environment_minister.

McKnight, D. 2005. 'Murdoch and the culture war', in R Manne (ed.), *Do not disturb: Is the media failing Australia?* Melbourne: Black Inc, pp. 53–74.

—— 2012. *Rupert Murdoch: An investigation of political power*. Sydney: Allen & Unwin.

McKnight, D. and Hobbs, M. 2017. 'Fighting for coal: Public relations and the campaigns against lower carbon pollution policies in Australia'. In B. Brevini and G. Murdock (eds), *Carbon capitalism and communication*, London: Macmillan, pp. 115–29.

Minerals Council of Australia. 2021. *Coal: Building Australia's future*, https://www.minerals.org.au/coal-building-australias-future.

Murphy, B. and McGee, J. 2018. 'Lawfare, standing and environmental discourse: Phronetic analysis', *University of Tasmania Law Review* 37(2): 131–68.

Nixon, C., Konkes, C., Lester, L. and Williams, K. 2021. Mediated visibility and public environmental litigation: The interplay between inside and outside court during environmental conflict in Australia. *Laws* 10(2): 35.

Preston, B. J. 2016. 'The contribution of the courts in tackling climate change', *Journal of Environmental Law* 28(1): 11–17.

Readfern, G. 2015. 'Time to examine the fossil fuel industry's relationship with Queensland's government', *The Guardian*, 9 October, https://www.theguardian.com/environment/planet-oz/2015/oct/09/time-to-examine-the-fossil-fuel-industrys-relationship-with-queenslands-government.

Robertson, J. 2015. 'Conservation group challenges approval of Carmichael coalmine as "illegal"', *The Guardian*, 9 November, https://www.theguardian.com/environment/2015/nov/09/conservation-group-challenges-approval-of-carmichael-coalmine-as-illegal.

—— 2016. 'Adani Carmichael coalmine faces new legal challenge from Conservation Foundation', *The Guardian*, 19 September, https://www.theguardian.com/business/2016/sep/19/adani-carmichael-coalmine-faces-new-legal-challenge-from-conservation-foundation.

Samuel, G. 2020. 'Independent review of the EPBC Act', *Commonwealth of Australia*, https://apo.org.au/sites/default/files/resource-files/2021-01/apo-nid310681.pdf.

Schliebs, M. 2016, 'Bid to halt Adani's Carmichael coal mine fails', *The Australian*, 29 August, https://www.theaustralian.com.au/business/mining-energy/bid-to-haltadani-carmichael-coal-mine-fails/news-story/06dfb23b53b83298ed6fb40e78c9d4b8.

Schuijers, L. and Young, M. A. 2021, Climate change litigation in Australia: Law and practice in the sunburnt country. In I. Alogna, C. Bakker and J.-P. Gauci (eds), *Climate change litigation: Global perspectives*. Leiden: Brill, pp. 47–78.

Seccombe, M. 2018. 'Rinehart's secret millions to the IPA', *The Saturday Paper*. 28 July–3 August, https://www.thesaturdaypaper.com.au/edition/2018/07/28.

Slezak, M. 2016a. 'Great Barrier Reef: New chapter opens in the fight to save natural wonder from mining', *The Guardian*, 10 April, https://www.theguardian.com/environment/2016/apr/10/great-barrier-reef-new-chapteropens-in-the-fight-to-save-natural-wonder-from-mining.

—— 2016b. 'Activists launch fresh court challenge over Carmichael coalmine', *The Guardian*, 3 May, https://www.theguardian.com/environment/2016/may/03/activists-launch-fresh-court-challenge-over-carmichael-coalmine.

—— 2020. 'Australia's Future Fund "in bed with Adani" after freedom of information request reveals $3.2 million investment', *The Guardian*, 16 December, https://www.abc.net.au/news/2020-12-16/future-fund-invests-millions-in-adani-project/12984734.

Smee, B. 2019. 'Adani mine would be "unviable" without $4.4 bn in subsidies, report finds', 29 August, https://www.theguardian.com/environment/2019/aug/29/adani-mine-would-be-unviable-without-44bn-in-subsidies-report-finds.

Stevens, M. 2016. 'Adani prepares for an end to lawfare with a smaller, cheaper Carmichael', *Australian Financial Review*, 21 September, https://www.afr.com/news/economy/adani-prepares-for-an-end-to-lawfare-with-a-smaller-cheaper-carmichael-20160921-grla4o.

Stutzer, R., Rinscheid, A., Oliveira, T. D., Loureiro, P. M., Kachi, A. and Duygan, M. 2021. 'Black coal, thin ice: The discursive legitimisation of Australian coal in the age of climate change', *Humanities and Social Sciences Communications*, 8(1): 1–9.

Tankard, J. 2003. 'The empirical approach to the study of framing', in R. Stephen, O. Gandy and A. Grant (eds), *Framing public life*. Mahwah, NJ: Lawrence Erlbaum, pp. 95–106.

van Vonderen, J. 2015. 'Australian Conservation Foundation challenges Adani's Carmichael coal mine in Federal Court', *ABC News*, 9 November, https://www.abc.net.au/news/2015-11-09/adani-mine-australian-conservation-foundation-court-challenge/6923598.

Wangan and Jagalingou Family Council. 2019. 'Latest news', https://wanganjagalingou.com.au/category/latest-news.

Wickham, G. and Goodie, J. A. 2013. *Legal and political challenges of governing the environment and climate change: Ruling nature*. London: Routledge.

Zhou, N. 2020. 'Crooked not courageous: Adani renames Australian group Bravus, mistaking it for "brave"', *The Guardian*, 5 November, https://www.theguardian.com/business/2020/nov/05/crooked-not-courageous-adani-renames-australian-group-bravus-mistaking-it-for-brave.

Dr Claire Konkes is a Senior Lecturer and Head of Discipline at The Media School, University of Tasmania. Her research continues to explore the role of media, especially news media, in the development of policy and law, especially in relation to gendered violence and the environment.

Dr Kathleen Williams is the Director of Creative Curriculum and a Senior Lecturer in Media at the University of Tasmania, Australia. She primarily researches the cultural, environmental and industrial impacts of changing media technologies and practices.

Dr Cynthia Nixon has a bachelor's degree in Environmental Engineering, a Master's in Environmental Law and a PhD in Media. Her research has focused on the intersection of strategic communication, litigation, media, and activism. She has worked for over 15 years in the energy and paper industries and is currently a consultant working to improve the sustainability of organisations by improving their environmental, social and governance (ESG) performance. Cynthia's current focus is supporting the rapidly growing renewable energy sector as the world transitions to a cleaner energy landscape.

Professor Libby Lester is Director of the Institute for Social Change and Professor of Journalism, Media and Communications at the University of Tasmania, Australia. She works to understand the place of public debate in local and global decision-making, and her research on environmental communication and conflict is published widely.

Urannah: The isolated home of rare species

Peter McCallum
peter@conservation.org.au

Figure 1
Urannah. Photograph credit: Jeff Tan.

Photographer Jeff Tan dropped into the Mackay Environment Centre back in 2015. He had been on an expedition to Urannah Creek, where he had the chance to photograph some delightful landscapes. Jeff showed me one of his photos, evocatively named 'Urannah_landscapes_24', which was taken as the sun set over the river. The deep shadows created an eerie, dark scene but, even in the dying light, the colours of the river rocks were easily visible through the clear, fast-flowing water. I wanted to learn more about the place.

Over the past six years, I have become involved in a campaign to protect the Urannah Creek and other rivers and streams that are threatened by a major new dam. Despite never visiting Urannah Creek in person, I feel a strong vicarious connection with the place. Traditional custodians have spoken to me about their deep connections with their sacred country and how dislocation occurred.

Bushwalkers have told me about their experiences of Urannah's ruggedness and unique beauty. Many biologists have shared knowledge about the river's ecological value.

Urannah Creek begins in the Eungella National Park, west of Mackay. On the eastern side, maritime trade winds blow moist air up against the steep escarpment. That's why the place is known as Eungella, which means Land of Clouds. It is also why dense tropical rainforest still exists there despite the gradual drying of the Australian continent over 180 million years. The rainforests are closely aligned with the Wet Tropics, which has led to calls for UNESCO World Heritage Listing by prominent scientists.

These rainforests were once home to the Northern Gastric Brooding Frog (*Rheobatrachus vitellinus*). Until the 1980s, there were two gastric brooding frog species. Both are now listed as extinct. In both species the female frog lays her eggs then immediately swallows them; her stomach acts like a uterus, where the eggs develop. Sometime later, fully formed frogs are birthed through her mouth. The Northern Gastric Brooding Frog was discovered in 1984 in the Eungella rainforest. By 1985, it was presumed extinct — most likely another victim of chytrid fungus. However, all hope is not lost for the species.

Dr Conrad Hoskin, a biologist based at James Cook University in Townsville, has a hypothesis that Urannah Creek may provide a suitably warm habitat for chytrid fungus to be ineffective. The critically endangered Armoured Mistfrog was also thought to be extinct until a small population was found in the west-facing splash zone of a waterfall in the Wet Tropics in 2008. Chytrid fungus is most effective when the frog's body temperature is 18 to 20°C. That's the temperature in shady, rainforest uplands. Chytrid is less effective in warm lowland streams. In 2021, Dr Hoskin was scheduled to lead an expedition to the Urannah Creek, which also contains rocky locations exposed to sunlight. There is a slim chance that the presumed extinct Northern Gastric Brooding Frog may still exist in Urannah Creek.

Urannah Creek is home to hundreds of species, including 30 that are listed as rare or threatened. One that is not on that list is Irwin's Turtle (*Elseya irwini*). As the fast flowing Urannah Creek tumbles over rocky rapids, it absorbs oxygen from the air. The Irwin's Turtle has evolved to take advantage of the oxygen-rich water. It can pump water through its cloaca (the multi-purpose opening for the digestive, urinary and reproductive tracts) and across respiratory filaments inside its body. As a result, female Irwin's Turtles can stay submerged for 90 minutes or more. They congregate near riffles where the oxygen concentration is highest. A dam will slow the flow of water, reducing the oxygen content. Consequently, the Irwin's Turtle will spend less time foraging and more time at the surface. This species, discovered only 30 years ago by famed naturalist Steve Irwin, could be pushed to extinction by a dam.

One of the great tools the Queensland Government provides citizens is the Queensland Globe. It is possible to take a virtual tour of Urannah Creek using 'The Globe'. If you trace your way along this stream from Mt Dalrymple, you will see that it consists of many waterholes and waterfalls, mostly surrounded by dense forest. A bushwalker recently told me he had made his way to Urannah Creek. After an arduous journey battling lawyer vine and leeches, he was rewarded by being able to watch a platypus swimming along the bottom of one of those deep waterholes, picking its way among the pebbles on the bottom.

Urannah Creek is part of a wetland aggregation that is listed on the Directory of Important Wetlands of Australia. The listing says it contains 'some of the best and least disturbed examples of riverine wetland occurring in Central Queensland'. Much of the system is undisturbed due to its isolation and ruggedness. That is part of the reason it is also considered a good place to build a dam. The steep country and constant flow of water mean a huge volume of water can be stored behind a relatively narrow dam wall. That clean rainforest water would be used to wash impurities out of coal to increase its economic value. It would also be used to create new areas of irrigated agriculture in catchments that flow to the Great Barrier Reef.

I am working with a team of passionate people at Mackay Conservation Group to protect Urannah Creek. We know that there are good alternatives to building a dam that will protect this landscape and provide enough water for farmers. Places like Urannah Creek should be protected for their intrinsic beauty and because there must be a place for nature in this world.

Peter McCallum is a jack-of-all-trades, having worked as a farm labourer, truck driver, graphic artist and event organiser among other jobs. Peter joined Mackay Conservation Group when he moved to the region in 1993. He is currently the group's coordinator.

Women of the Great Barrier Reef: Stories of gender and conservation

Kerrie Foxwell-Norton, Deb Anderson and Anne M. Leitch

K.Foxwell@griffith.edu.au
Deb.Anderson@monash.edu and
anne.leitch@csiro.au

Abstract

In the late 1970s, Carden Wallace was at the beginning of her lifelong exploration of the Great Barrier Reef — and indeed, reefs all over the world. For Wallace, who is now Emeritus Principal Scientist at Queensland Museum, the beginning of her Reef career coincided with the emergence of both feminist and environmental movements that meant her personal and professional lives would be entwined with a changing social, cultural and political milieu. In this article, we couple the story of Wallace's personal life and her arrival in coral science to identify the Reef as a gendered space ripe to explore both feminist and conservation politics. The article is part of a broader Women of the Reef project that supports a history of women's contribution to the care and conservation of the Reef since the 1960s. In amplifying the role of women in the story of the Reef, we find hope in the richness of detail offered by oral history to illuminate the ways discourse on the Reef and its women sits at the intersection of biography, culture, politics and place. In these stories, we recognise women's participation and leadership as critical to past challenges, and to current and future climate change action. By retelling modern Reef history through the experiences and achievements of women, we can develop new understandings of the Reef that disrupt the existing dominance of patriarchal and Western systems of knowledge and power that have led us to the brink of ecological collapse.

The lovers can see, and the knowledgeable. Annie Dillard[1]

Carden Wallace fondly recalls being there that exciting night, when the moon was right and scientists first witnessed the mass spawning of coral in 1981. 'The first year that we saw coral spawning,' Wallace says, 'we were on Magnetic Island and everybody had sort of predicted when it was going to happen. But it kept *not* happening.' Having followed the clues, but been made to wait, most of the team of researchers had dispersed. 'Some of the people had to go home,' she recalls. Yet Wallace and one other James Cook University (JCU) scientist opted to stay behind on the beach that night.

I had my kids with me at that time. So I'm sitting on the beach in the middle of the night and he [the other scientist] was just snorkelling ... it's quite shallow ... then he stands up and is making all these noises through his snorkel, screaming out. And they were all spawning, and everything was ... spawning. Just going off ... I had to wait until he came back up and say, 'Don't you leave my kids!' [she chuckles] And I had to go in by myself.

The underwater snowstorm off the coast of Townsville, the natural wonder of coral polyps simultaneously releasing egg and sperm bundles for external fertilisation, garnered global attention. It graced the cover of the esteemed journal *Science* in 1984. It stimulated an extraordinary effort to document spawning times in other parts of the globe and significantly advanced knowledge on coral reproduction, itself fundamental to the science that can help reefs survive. For their pivotal discovery, the team of JCU scientists received the prestigious Eureka Prize for Environmental Research in 1992. There is a certain irony (or perhaps injustice) in the way this spectacular beginning-of-life moment in one of the most immense, impressive, living non-human ecologies on the planet is officially reported and remembered, narrowing human life to (only) its success in capturing said nature's event. At the very moment we witness and celebrate the spectacle of Reef life, we curtail our appreciation of the human lives and experience that enabled such a moment in the first instance. This curtailment has specific consequences in women's lives — and the way their lives are obscured in the historical record — but also, more broadly, for how we define human life and our relations to the Great Barrier Reef. In the official recall of a critical moment in the history of the Reef and coral reef science, we affirm the separation of public lives from private — of humans from nature, scientific from aesthetic, reason from emotion and the raft of dualisms characteristic to modernity. The tapestry of human lived experience epitomised in the entwining of one woman's private life and professional achievement remains unseen at precisely the moment the magnificence of Reef life is celebrated.

***

Four decades on, mass coral spawning could be read as a mighty demonstration of persistence, given that mass coral bleaching due to climate change has become an escalating existential threat. Certainly, with corals today 'on the front lines of the Anthropocene', as ethnographer Irus Braverman writes in *Coral Whisperers*, the scientists who study them have found themselves in the vanguard of conservation science.[2] This description refers to coral conservation champions like Carden Wallace,[3] Emeritus Principal Scientist at the Queensland Museum. Importantly, it speaks also to the culture and society that has inflected the questions that scientists have asked of reef systems through the emerging ecological tragedy and its response. Momentous gains have been made in scientific knowledge of the Great Barrier Reef over the past 60 years, in part because scientists have been galvanised by mounting social, political and ecological threats to the 3,000 or so individual reef systems therein. This period of stress stretches from the first major outbreak of the crown-of-thorns starfish at Green Island in the 1960s,[4] through the '1965–75 war' to save the Reef from the lethal threats of limestone mining and oil drilling,[5] to the back-to-back mass bleaching events that wiped out half of the Reef's shallow corals in 2016–17[6] and severely depleted coral larvae. With bleaching in 2020 deemed the most widespread ever recorded,[7] coral scientists have declared a climate emergency.[8] Almost every scientist now working on reefs is working to save them. They are driven to find ways forward; as Braverman writes, they have to empower themselves 'to continue their work despite the daunting tasks and the low odds of success'.[9]

In this period of immense changes and oscillations between hope and despair, spare a thought for the 'woman in a man's world'. We borrow the latter term from Wallace in a bid to expand our repertoire of responses to reef loss and climate crisis — to keep on finding new ways forward. As feminist Anne Summers[10] notes, a 'profound ignorance' about the roles of women has pervaded Australian history, leading to a suppression of their contribution in the telling of the nation's history. We note, for example, that women have had to battle for inclusion and fair representation in those domains of politics and science where the official story of the Reef has largely been told. Precisely because of their struggle for social equality, women's historical influence upon and contributions to Reef conservation have been overshadowed or ignored. In counterpoint, this article offers insights from a larger project that aims to uncover and share the untold stories of women who have, since the 1960s, championed the value of the Reef to Australians and others abroad.[11] We intend to recover an important chapter of Queensland's past and to fill a notable gap in the historical record. Our preliminary research has revealed that women of many walks of life have made landmark contributions to Reef knowledge, regulation and protection during this period, participating both as individuals and through collective action in urban, regional and remote parts of the state. Their interventions range from the crucial voices of First Nations women to the endeavours of female researchers in a variety of disciplines; from the remarkable efforts of environmental activists, artists and journalists to the progressive struggles of women in business, public service and the unions — and more. By retelling modern Reef history through women's experiences and achievements, we seek to develop new understandings of the Reef that disrupt and interrogate the existing dominance of patriarchal and Western systems of knowledge and power. We take as our inspiration and starting point the final words of ecofeminist scholar Val Plumwood in her seminal book, *Feminism and the Mastery of Nature*:

> If we are to survive into a liveable future, we must take into our own hands the power to create, restore and explore different stories, with new main characters, better plots, and at least the possibility of some happy endings.[12]

We intend to build upon feminist scholarship that sheds critical light on the intersection of two twentieth-century political movements, women's rights and ecological sustainability, in order to highlight the interconnectedness of social and environmental injustices.[13] In *Feminist Ecologies*, a ground-breaking anthology of research conducted in the Australian and New Zealand context, Lara Stevens, Peta Tait and Denise Varney chart the particular significance and struggles of ecofeminism in challenging a range of oppressive structures. Arguing that feminist thought remains 'vital if we are to redress the near universal neglect of women by cultures around the world, including academic ones',[14] they show how Australian and New Zealand scholars and activists have developed what ecofeminist Carolyn Merchant terms 'an ecofeminist ethics of earthcare',[15] which interrogates the particular ways our colonial pasts have degraded the environment. Historian Emma Shortis has similarly explored the contribution of women's activism against mining in Antarctica, as well as women's long history of living on the continent. Having travelled initially as companions to their explorer husbands, women later came as independent scientists and activists in their own right.[16] Work of this kind also underlines the connection between the local and global: by calling for a deeper understanding of women's engagement in one site, it simultaneously illuminates their 'larger role in the international environmental movement'.[17] This local–global nexus is particularly important, argue feminist scholars Stevens, Tait and Varney, 'given the intersecting nature of climate and/or rapid environmental change and feminist concerns'.[18]

This crucial intersection is both the backdrop and the inspiration for the story of Carden Wallace that informs this article. Hers is among the twelve oral history narratives recorded so far for the project. Each presents distinct conceptions of the Reef, different views of the gendered relationship between human and non-human nature and poignant reflections on the significance of broader national and international movements for women and environments. Carden Wallace's story is emblematic of the evolving relationship between conservation and science in Queensland. In 1970, long before she was pacing the beach at Magnetic Island, Wallace had been perched in the rammed-earth basement of the Queensland Museum prising open scores of ammunition boxes to reveal hessian-wrapped coral skeletons from Low Isles collected by Australian scientist Charles Hedley in the years leading up to the British Expedition to the Reef of 1928–29.[19] (Wallace is unapologetic: 'It was like going to heaven.') She was also among the experts in 2008 at Bikini Atoll, five decades after American nuclear testing in the Pacific, in order to report on the capacity for coral biodiversity recovery after extreme disruption.[20] And Wallace is here, still marshalling research priorities, looking long and hard at how we manage, conserve and understand coral reefs amid the world's sixth great extinction.[21]

By tracing aspects of Wallace's story, this article illuminates the challenges faced by female Reef researchers while working in spaces, both real and imagined, that were and remain dominated by men. It also underscores the fact that science is a situated practice, one that is both historically specific and tied to knowledge and power, as well as contingent on the social and material forces that circumscribe possibility and action. This article, then, is less about women in science *per se* and more concerned with illuminating Reef history as a gendered space. Our decision to explore a single narrative is thus pragmatic, allowing room for the richness of detail offered by oral history to illuminate the ways discourse on the Reef and its women sits at the intersection of biography, culture, politics and place.

***

Australia's identity is frequently and publicly embedded in the idea of nature and the nation's love for it. Indicative is historian Mark McKenna's reflection on the numerous attempts by notable Australian literary and public figures to rewrite the constitutional preamble in ways that define the land through 'its poetics of place, as a source and inspiration for national unity'.[22] Being an inland continent where relations to coastal and marine environments are defining for a population that lives largely by the coast, the Great Barrier Reef constitutes the marine equivalent of this national love affair with nature.[23] Of course, the idea of an 'Australian culture of nature' is politically charged, being both inclusive and exclusive by degrees, and translates across time and place. Yet there are serious exclusions when the Reef is understood and communicated primarily as a scientific object, or as an economic asset via the locus of tourism or fishing industries.[24] Other important ways by which we communicate the value of the Reef and its connectedness to us are thereby diminished by the power of industry, politics and science to commandeer priorities and debates. Power, privilege and knowledge characteristic of those domains can also generate many tensions, including — as in Wallace's story — those between the public lives of men and the private lives of women.

In grounding new knowledge on the Reef in women's voices, we therefore start here. We cannot write about or celebrate the achievements of women in public life without also considering the private sphere.[25] In this respect, and as scholars of feminist biography note, we face a fundamentally different task to the biographers of men. For example, this may be necessary in order to explain why and how our subjects have stepped outside of the private sphere, or we may need to uncover the remainder of the story that is missing in men's biography, which celebrates men's public achievements and contributions yet pays little attention to private lives and to the roles of women within them. This includes the private support of women at home *and* in research, and how men have maintained positions of dominance. Here we note the power and application of oral history, which has already challenged the historical enterprise, if not the hegemony of scholarly authority, generating heated debate over the relationship between memory and history, past and present.[26] As historian Joan Sangster argues, the feminist embrace of oral history has its roots in a recognition that traditional sources have often neglected the lives of women, offering scholars a means of 'even contesting the reigning definitions of social, economic and political importance that obscured women's lives'.[27]

Our project uses 'life story' interviews both as a way of putting women's voices, stories and experiences at the centre of history and of deploying gender as a category of analysis. Interviews conducted so far with identified 'women of the Reef' have been structured to focus explicitly on women and their lives, roles and contributions, a process that inevitably shapes the respondents' narrative response and interpretive angle. Insofar as this work of environmental scholarship intersects with feminist biography, it is not simply about uncovering the lost lives of women and their notable contributions with respect to the Reef. Instead, our focus is on gender, on gaining insight into the mechanisms that have worked to perpetuate a status quo that has always favoured men, and on revealing relatedly dominant ways of knowing nature that exclude or marginalise others such as women and First Nations people, as well as the intersections between them.

***

It was at the Palm Island exhibit in Brisbane's annual Royal Queensland Show, more commonly called the Ekka, and its mid-twentieth century 'phantasmagoria' of human abundance, that a young Carden Wallace first laid eyes on corals.[28] 'Our grandmother used to take us every day,' she recalls of the sea of Ekka stalls showcasing regions of the state and the Palm Island exhibit that captured her imagination.[29] 'There were people from Palm Island there, and they had all these corals that they'd painted, and you could buy tiny little ones.' A girl from suburban Brisbane had thus just glimpsed the aquatic marvel that is exquisitely not human and whose revelation and salvation would become her life's work. Wallace calls her first encounter with corals 'a bit weird'. But we insist that it was wonderfully prophetic and poetic that her first encounter with corals should have taken place through a meeting with the Bwgcolman mob from Palm Island, First Nations and Traditional Owners of the reefs off Townsville, where she would live for thirty years as a marine scientist. The encounter with these women is an ordinary instance of intersecting histories of women who understand the Great Barrier Reef in ways that are nevertheless distinctive. First Nations women and their relationship to Country constitute the dimension of human–nature relations that begins our story of being here and how we have fought for recognition in formal corridors of power as legitimate and authoritative holders of Reef knowledge.[30] Aileen Moreton-Robinson's *Talkin' Up to the White Women* is a reminder that we must continually interrogate our own subject positions and navigate the subjugated knowledges, spiritualities and experiences of First Nations women and the relation of whiteness therein.[31] Wallace's chance encounter points to the unspoken and unavoidably shared, yet distinctive, heritage of our women of the Reef.

Wallace can identify strong female figures in her upbringing in the southern Brisbane suburb of Moorooka, where her grandmother inspired an early interest in nature and her mother offered a model of self-determination and empowerment through education. Both Wallace's parents had worked. Her father was a detective with the Criminal Investigation Branch; her mother went nursing and then wanted to go to university. Given that the mid-century education system and labour force were characterised by a marked division of the sexes and their expected roles in society,[32] this meant her mother having to put herself through university — in this case, matriculation and tertiary study right through to a PhD — in order to become a clinical psychologist. 'Incredible,' Wallace says of the effort required to act outside the domestic structures and the conservative norms of post-war Australia. 'She spent most of my childhood going to night school,' she recalls. 'There was nothing about it that was easy.'

Greater educational opportunities came earlier in life for Wallace, as well as plenty of 'luck'. Intriguingly, it was the nuns at the all-girls St Ursula's College, today the site of the Brisbane Catholic Education Centre, who opened the door to the world of science.[33] She recalls their support for 'a good education' and a curriculum that she presumed was not the norm in the early 1960s. 'I don't think a lot of schools really had zoology and physiology and courses like that, but we did.' She remembers her excitement at entering the school's chemistry room, stacked with equipment and with raked seating 'just like a tiny imitation of a university room'. There began her fascination with marine life, particularly invertebrates. 'I just loved hearing about things I'd never heard about before.' Only a minority of St Ursula students made it into the lab, however. Gendered expectations of study, work and roles in society literally divided the student body. Of Wallace's Year 9 cohort, she says, some 100 girls were 'hived off' for commercial studies; the remaining 20 or so went into the academic stream. She views her consignment to the latter to be a stroke of luck. 'Sometimes I think maybe we weren't pretty enough,' she jokes, but she also admits that she is still grappling with the absurdity of the division and its impact on women's career paths. 'I get together sometimes with all these women and there's no reason why one lot should be considered any cleverer than the other lot. We were just lucky.'

***

Wallace studied hard in her teens to gain a Commonwealth scholarship and entry to a Bachelor of Science degree at the University of Queensland, which was followed by First Class Honours. Amid the expansion of tertiary education in Australia in the

1960s and the increase in scholarships for universities and teachers' colleges, which granted new opportunities to young women as well as men,[34] she was among the growing number of Queensland women entering the fields of science, technology, engineering and mathematics (the STEM disciplines). But she says 'the promise' for women with the ability and desire to succeed in male-dominated domains 'didn't play out for everybody':

> It was a time of — well, it was a bit of a trick really. Because you went to uni, you thought *Ooh, there's women all around* and *This is our generation, This is what we're going to do*. But it's only later in life you realise that so many people just got married and ... [gave] up whatever it was that they were studying for.

But the dream was ignited. As historian Elaine Martin writes, new opportunities in the 1950s and 1960s brought aspirations for life experiences beyond the home, and although many of these Australian women still married young and had children while young, 'what did not show up in the census data were their dreams for their future lives'.[35]

An impressive array of recent feminist scholarship has highlighted the relationship between scientific knowledge and gender inequalities, interrogating the strictly guarded sociocultural boundaries around membership in the science community.[36] Even today, increased women's participation and advancement in science have yet to translate into gender equity in the Australian workforce: the research data show that only 16 per cent of the nation's STEM-skilled workforce are women.[37] Women researchers remain squeezed out of science careers by critical structural barriers and key factors of organisational culture.[38] The Australian Academy of Science points to the male-dominated leadership of research institutions, the 'outdated and embedded' views of women academics and the raft of career progression and promotion processes that fail to account for the career breaks women may have to use to care for family.[39] Another noted obstruction is the persistence of an organisational culture that allows — even validates — sexual harassment.[40] This is not to say that Wallace hasn't witnessed or experienced such barriers. Her pathway through the sometimes-hostile terrain of science — and her tendency (with us at least) to speak less of competition or aggression and more of connection and cooperation with other women we might call 'women of the Reef' — is all the more remarkable given what she has had to overcome. Indeed the stories of women like Wallace, who matured during the 1960s and 1970s counter-movements, offer significant insight into what feminist historian Lynn Abrams terms 'a bigger story about the progress of women in the post-war era',[41] as well as the rising environmental consciousness of the time. The context for the early years of Wallace's career is marked by its challenge to existing orthodoxies surrounding gender and nature, and for the emergent and ongoing challenge to Western definitions of both.

On reflection, Wallace says, while there was a notable female presence on the university's St Lucia campus in the 1960s, few were studying geology, even though the research professor of geology was none other than Dorothy Hill. In the decade prior, Hill had become the first female professor at an Australian university. She was a former secretary of the Great Barrier Reef Committee (1946–55) and later the first female president of the Australian Academy of Science (1970). Her specialty was fossil corals. Wallace recalls that Hill was known as 'not too kind to girls',[42] yet her connection with this well-known woman researcher benefited her career: 'I had the good luck to have a boss who was one of her favourite boys,' Wallace says, 'and she sort of took me under her wing a little bit.' This helped her gain references when applying for grants to fund her research.

Wallace acknowledges that her research career began during an extraordinary period in Queensland for activism and change, both in environmental and social terms. Concomitantly, it was a significant period for the advancement of marine science on the Reef. By the mid-1960s, as historian Iain McCalman points out, there were just three basic marine research stations on the Reef and none of these employed full-time research scientists.[43] Wallace was at the University of Queensland when the first outbreak of crown-of-thorns starfish (COTS), which prey upon hard or stony coral polyps, gripped the field. The COTS presence on the Reef was believed to be a consequence of the commercial and tourist removal of its chief predator, the Triton shell.[44] 'That was very, *very* dramatic,' Wallace says, likening COTS to COVID-19 in the sense that 'nobody knew what was going to happen next'. Some scientists feared 'whole reefs would get eaten out'.

Meanwhile, under the leadership of pro-development premier Joh Bjelke-Petersen, the Queensland Government was determined to establish an offshore petroleum industry and by 1968 had opened up the state's coastline to oil exploration.[45] McCalman notes that a 'persistent problem' faced by the network of individuals like Judith Wright and the associated organisations and events that emerged in the 1960s to campaign on behalf of the Reef was the lack of Australian scientists qualified in Reef coral biology and ecology. Wallace says the Queensland Littoral Society (formed in 1965 as the precursor to the Australian Marine Conservation Society) began training university zoology students to dive by taking them on research trips to North Queensland. 'A lot of us learnt to dive then,' she says. In 1969, the state and federal governments established a joint inquiry into potential damage to the Reef. Wallace remembers sitting in court, listening to scientists speak. She still marvels at the outcome, with the Great Barrier Reef Marine Park, established in 1975, saving the reef from oil drilling and mineral extraction.

Wallace herself began diving not in North Queensland but on the corals of Moreton Bay and Heron Island, where Hill had been instrumental in establishing the Reef's first research station in 1951. Wallace landed her first curatorial position in her early twenties. 'I think I always wanted to work at the Queensland Museum,' she says. 'A job came up when I was just finishing my Honours.' She remembers finding it to be a relatively progressive museum workplace for women in 1970:

> There weren't any women in high places at the Museum, but quite a lot of women worked there, and they weren't really averse to employing women as curators. There were about four or five women curators when I was there, and more later on. So that was quite a difference from most of the other museums in Australia.

This contrasts sharply with her recollection of early days 'in the field' in North Queensland, when she gained an intimacy with the rituals of the mariner's life.

> When I first started organising field trips to the Great Barrier Reef, I was living in Brisbane. I used to have to drive up to Townsville in the Museum Falcon or something *(chuckles)* and go onto the wharves and try and find somebody whose boat we could go out on. And so: it was bad luck to have a woman on your boat ...

Wallace can laugh now at the formative experience of being a twenty-something scientist viewed with suspicion, of having to drive 1,300-odd kilometres up the Queensland coast to encounter the Reef and find her gender a liability, and of her relief on finding two Cardwell local skippers who were also brothers, who 'didn't mind' having a female researcher on board their big fishing boats. Yet the impact of being an invader within this aquatic realm was made all too clear. 'That's the basis that we had to start from,' she says. 'It is an example of how difficult it was to organise *any* research, really — if you were a woman in a man's world.'

* * *

In the mid-1970s, during her seven years of working with the Queensland Museum, Wallace was awarded a major grant from the Advisory Committee on Research on Crown-of-Thorns Starfish that took her abroad. At the time, it was thought that staghorn corals (*Acropora* species) — common, fast-growing corals named for their branching shape, which provide reefs with much colour and beauty — were the most attractive to the crown-of-thorns predator. Even so, scientists had yet to sort out the nomenclature of these corals, Wallace says. She set about the task of identifying type specimens, the specific specimens of coral that serve as a reference point when a species is first given a scientific name. This entailed comparing what was held in the Queensland Museum with type material held in the main museums around the world. 'I packed up for myself a polystyrene suitcase, with little holes in it for a specimen of each of the species that I needed to confirm,' she says, and off she went overseas to visit (among others) the Smithsonian, the British Museum, the Paris, Netherlands and Berlin Museums and the National Museum of the Philippines. 'There was an American expedition that went round the Pacific in the mid-1800s; they had this *beautiful* collection of types, and I spent a lot of time trying to identify my specimens against those …'

While interviewing Wallace, moments like this prompted a line of questioning that, on reflection, moved from her individual account to the consideration of wider social factors. When we raised with Wallace that global museum-trotting in the mid-1970s would have been slow, given the challenge of patching together many long-haul flights and land transport logistics, she agreed: 'It was quite an adventure really.' This was one such moment when her achievements in public life could not be fully apprehended without also considering the private sphere. We asked:

So how many children did you have?

Well, I had two, but I had one when I was doing that, and he was about nine months old.

And you took him around the world with you?

Well, no, I left him … My mother and my mother-in-law and my husband, and a lady who looked after him during the day, they shared it all.

How long were you gone for?

About six weeks, I think … But beforehand I made up all these little meals for him and had them in the freezer … every meal for him …

Was it hard for you ... was it tough for you to leave him? How did you negotiate that? 'Cause, you know, we've all got children.

Yes, well, I think – so I had the grant to do it, and all those people were encouraging me to do it ... It was wonderful that they were willing to do it, you know? I missed him the whole time. It was terrible really! But I don't think he would have missed me too much because he had all these lovely people swapping him around!

But to be encouraged by all those people in your life to say, 'We've got this. You go!' ...

Mmm. One of my daughters-in-law was horrified when she found out. She said to me, 'Who leaves their baby and goes overseas?' And I said, 'Well, I did.'

Here we need to acknowledge our agency as interviewers informed by the belief that we must both celebrate women's achievements and lay political claim to women's experiences of oppression. As academics cognisant of the myriad ways in which women's lives are impacted by a patriarchy, this women-as-mothers moment offered an apt lens through which to understand how Wallace has had to juggle her private and public lives. It throws into relief the intersecting complex of political-economic, sociocultural and material-environmental processes required of a woman scientist in order to maintain everyday life and to sustain human cultures and communities across time scales that range from the daily to the cross-generational. Wallace's fortune was to have private champions who supported her challenge to gender norms and enabled her to continue in her work. For others of her generation, this 'motherhood penalty' — both biologically determined and culturally constructed — meant professional careers were either willingly or unwillingly abandoned or became too difficult to juggle with children.[46] Wallace was able to overcome such motherhood impediments by having a mother who set an early example through her prolonged period of dedication to her own career. Later, Wallace and her mother travelled to work together, another way of diminishing the distance between the private lives of mothers and children and the public lives of women and work.

In 1979, when Wallace finished her PhD at the University of Queensland — on staghorn corals and their distribution on the Great Barrier Reef — she was offered a job by John 'Charlie' Veron, a scientist often referred to as the 'godfather of coral', having classified about 20 percent of the world's corals.[47] Working together at the newly established Australian Institute of Marine Science in Townsville through the early 1980s, Veron and Wallace published the *Scleractinia of Eastern Australia Part V (1984)*,[48] on staghorn corals. Fifteen years later, Wallace reviewed this family in *Staghorn Corals of the World: A Revision of the Genus* Acropora. As the first review of *Acropora* in more than a century, the book drew together Wallace's work on coral fossils and her vast experience of live corals, reminding us why we should care about reefs and the corals that build them.[49] In this book, her ability to straddle heritage and science is manifest, studying dead and live *Acropora* in a life's work that established her globally as the best-qualified to tell their story.

While Wallace is recognised with individual achievement, including being the first to describe a number of corals, she acknowledges the important collegiate and social networks that underpin a successful science career. She stresses the 'camaraderie and communication' she experienced 'among people who work on coral reefs and also people who work in museums'. Never far from her story is the support of others she has encountered along the way, including fellow women of the Reef Isobel Bennett, Patricia Mather, Bette Willis and Di Tarte — whose stories are also being gathered in our project. The presence of these important others in her story amplifies the social ecology of Reef science and reveals the lasting impact of 'a lot of very inspiring people that really covered certain bits of ground that needed to be covered in terms of how you go about living as a research scientist'. She credits these relations, often between women or those that supported her work personally or professionally, as critical to her productivity, innovation and extraordinary contribution to our knowledge and understanding of the Reef.

***

This article touches upon one woman's contribution to fostering the knowledge and protection of the Reef. Her work maps onto the biggest challenges, names and triumphs in coral science, having built a career that has spanned half a century, encompassed many of the world's great museums and investigated reef systems too numerous to count. Yet Carden Wallace doesn't view herself a 'woman of the Great Barrier Reef'. She could in fact be considered a woman-of-the-reefs *plural*, given her extensive experience studying and diving on reef ecosystems around the globe. Still, she demurs. 'That's the sort of thing that other people think of,' she laughs. 'I would think of myself as somebody who loves coral reefs.' So it is that she deploys the language of love, then a wistful nostalgia and lament, to describe 'the state of things' off the Queensland coast:

> I've dived on hundreds and hundreds of reefs ... Each dive seems to be — each experience seems to be — better than the last, you know? ... But I don't know if I'd want to go to too many damaged reefs or up and down the Great Barrier Reef and just see how dead they all are.

Actually, Wallace is well accustomed to research endeavours that oscillate between the living and the dead. Her very first job with the museum offered unique insight into reef expeditions that dated back through the 1800s, when ships travelled to Queensland waters to study the reef systems, either while they were exposed by the tides or by dragging up the coral. 'I guess they saw them alive,' Wallace ponders, 'just about to die.' In the contemporary context of climate emergency, however, where the life and 'death' of the Great Barrier Reef has become a torrent of politics and public debate about climate change and its impacts, Wallace continues to push the boundaries of what she and others know of the Reef. She sees hope in the research being conducted on the mesophotic zone — the low-light zone that sits below the brightly lit layer with which we are most familiar and the deep, blue depths. She believes that it may provide a safe haven for some species and a common lineage that could provide opportunities for others. She sees 'no barriers in the sea' in the sense that the Reef will be best understood in its wider context by gaining experience of the many reef ecologies around the globe. And she continues to see opportunities to learn something new.

We too see opportunities. By highlighting the contribution of women to Reef knowledge and conservation, and elevating women's participation and leadership as critical to past challenges, we aim to expand the repertoire of responses to climate crisis, and to current and future climate change action. Our work seeks to reinstate women in a historical record that is unbound by the structural domains of politics or science and open to the richness and variety of human connections with the Reef. It also represents a deeper challenge to the dualisms characteristic of modernity that separate reason from nature, and that distance humans from each other and from ecologies. It traverses ideas on the power to intervene, to free women from beneath patriarchal structures and to disrupt an appreciation of nature that would, as Plumwood wrote, prioritise reason and rationality over emotion, public lives over private, 'men over women, Europeans over barbarians and the uncivilised',[50] and, in doing so, diminish the rich kaleidoscope of experiences that engender love of the Reef.

This is the core work of the humanities, arts and social sciences (HASS) and their projects begun in history and heritage,[51] art and communication,[52] politics, sociology and the like, where human emotion and feeling, culture and meaning take centre stage in the bid to understand what the Reef means, and how and why it means so much to Australians and the world. For if the fate of the Reef lies in our hands, then we must accept that it encompasses meanings far wider and deeper than its often-cited extraordinary biodiversity, its role in planetary health and other scientifically defined wonders. This nature superstar is also a premier site of cultural imagination, a symbol to the world of who and what is 'Australian'. In a nation where nature is writ large, how we communicate our feelings for the Reef also matters crucially to how we remember our past and imagine our future. This is a story peculiar to Australia and our unique sense of place, of place meanings and of place changes,[53] but it is also salient across the world. Our relations to nature and environments are connected both to how we value or understand these places and to how we understand and manage relations between ourselves. The Great Barrier Reef is undeniably a consequential site of local and global significance through which to unpack the redemptive politics of gender and nature. And for our women of the Reef, the politics of gender are the waters in which they swim.

Acknowledgements

The authors thank the State Library of Queensland 2020 John Oxley Library Fellowship for funding this research. Special thanks to Dr Carden Wallace for her time and generosity and to our other women of the Reef for their support and kindness.

Notes

1 Annie Dillard, 'Seeing', *The abundance* (Edinburgh: Canongate Books, 2016), p. 157.
2 Irus Braverman, *Coral whisperers: Scientists on the brink* (Berkeley, CA: University of California Press, 2018), p. 6.
3 Expedition cruise company Aurora Expeditions in March 2021 dedicated a ship to five of the world's leading women conservationists. 'Coral conservation champion' Wallace was honoured with deck 6. In: Holly Payne, 'Aurora Expeditions' Sylvia Earle lauds

female conservationists,' *Seatrade Cruise News*, 22 April 2021, https://www.seatrade-cruise.com/shipbuilding-refurb-equipment/aurora-expeditions-sylvia-earle-lauds-female-conservationists.

4 Robert Endean, 'Population explosions of *Acanthaster planci* and associated destruction of hermatypic corals in the Indo-West Pacific region', in O. Jones and R. Endean (eds), *Biology and geology of coral reefs*, vol. 11, Biology I. (New York: Academic Press, 1973), pp. 389–438.

5 Judith Wright, *The coral battleground* (Melbourne: Thomas Nelson, 1977).

6 Terry Hughes, James Kerry, Andrew Baird, et al., 'Global warming impairs stock-recruitment dynamics of corals', *Nature* 568 (2019), 387–90.

7 Terry Hughes and Morgan Pratchett, 'We just spent two weeks surveying the Great Barrier Reef. What we saw was an utter tragedy', *The Conversation*, 7 April 2020, https://theconversation.com/we-just-spent-two-weeks-surveying-the-great-barrier-reef-what-we-saw-was-an-utter-tragedy-135197.

8 Michael Slezak, 'Climate emergency declared by 11,000 scientists worldwide who warn of "catastrophic threat" to humanity', *ABC News*, 6 November 2019, https://www.abc.net.au/news/2019-11-06/climate-change-emergency-11000-scientists-sign-petition/11672776.

9 Braverman, *Coral whisperers*, p. 248.

10 Anne Summers, *Damned whores and God's police* (Ringwood: Penguin, 1994).

11 This article presents oral histories gathered as part of a collaborative research project funded by a State Library of Queensland John Oxley Library Fellowship 2020, titled 'The Women of the Great Barrier Reef: Discovering the Untold Stories of Environmental Conservation in Queensland'.

12 Val Plumwood, *Feminism and the mastery of Nature* (London: Routledge, 2003), p. 196.

13 See, for example, Sherilyn MacGregor (ed.), *Routledge handbook of gender and environment* (London: Earthscan, 2017); Lara Stevens, Peta Tait and Denise Varney (eds), 'Feminist ecologies: Changing environments in the Anthropocene' (London: Palgrave Macmillan, 2018).

14 Stevens, Tait and Varney, 'Introduction', pp. 1–22.

15 Carolyn Merchant, Earthcare: Women and the environment (New York: Routledge, 1996), p. 186.

16 Emma Shortis, '"In the interest of all mankind": Women and the environmental protection of Antarctica,' in Stevens, Tait and Varney (eds), *Feminist ecologies*, pp. 247–61.

17 Shortis, '"In the interest of all mankind"', p. 247.

18 Stevens, Tait and Varney (2018), 'Introduction', p. 8.

19 Trisha Fielding, 'Expedition to the Great Barrier Reef 1928-1929 – Part 1,' James Cook University Library News, 14 August 2018. Available at: https://jculibrarynews.blogspot.com/2018/08/expedition-to-great-barrier-reef-1928.html.

20 Zoe Richards, Maria Beger, Silvia Pinca and Carden Wallace, 'Bikini Atoll coral biodiversity resilience five decades after nuclear testing', *Marine Pollution Bulletin* 56 (2008), 503–15.

21 Maria Beger, Russ Babcock, David Booth, ... John Pandolfi, 'Research challenges to improve the management and conservation of subtropical reefs to tackle climate change threats', *Ecological Management and Restoration* 12 (2011), e7–e10.

22 See https://www.griffithreview.com/articles/poetics-of-place.

23 Philip Drew, 'The coast dwellers: Australians living on the edge' (Ringwood: Penguin, 1994); David Booth, 'Australian beach cultures: The history of sun, sand and surf' (New York: Routledge, 2001); Leone Huntsman, 'Sand in our souls: The beach in Australian history' (Melbourne: Melbourne University Publishing, 2001); Jeremy Goldberg, Nadine Marshall, Alistair Birtles, ... Bernard Visperas. 'Climate change, the Great Barrier Reef and the response of Australians', *Palgrave Communications* 2 (2016), 15046.

24 Deloitte Access Economics, *At what price? The economic, social and icon value of the Great Barrier Reef* (2017), https://www2.deloitte.com/content/dam/Deloitte/au/Documents/Economics/deloitte-au-economics-great-barrier-reef-230617.pdf.

25 Rosemary Auchmuty and Erika Rackley, 'Feminist legal biography: A model for all legal life stories', *The Journal of Legal History* 41 (2020), 186–211.

26 Rob Perks and Alistair Thomson, 'Introduction to Second Edition', in Joan Sangster (ed.), *The oral history reader* (London: Routledge, 2006), pp. ix–xiv.

27 Sangster, *The oral history reader*, pp. 87–100.

28 Joanne Scott and Ross Laurie, 'Queensland in miniature: The Brisbane Exhibition', *Queensland Historical Atlas*, https://www.qhatlas.com.au/content/queensland-miniature-brisbane-exhibition.

29 Unless otherwise stated, quotes by Wallace presented in this paper are from Carden Wallace (2020), interview by Kerrie Foxwell-Norton and Anne Leitch, transcript, 'Women of the Great Barrier Reef: Oral History Collection 2020–21', Griffith University.

30 Diane Jarvis, Rosemary Hill, Rachel Buissereth, ... Wren, L. 2019, *Monitoring the Indigenous heritage within the Reef 2050 Integrated Monitoring and Reporting Program: Final Report of the Indigenous Heritage Expert Group*. Townsville: Great Barrier Reef Marine Park Authority.

31 Aileen Moreton-Robinson, *'Talkin' up to the white women': Indigenous women and feminism* (Brisbane: University of Queensland Press, 2000).

32 Australian Bureau of Statistics, *Australian social trends: 50 years of labour force statistics, now and then*, Canberra: ABS, https://www.abs.gov.au/AUSSTATS/abs@.nsf/Lookup/4102.0Main+Features30Dec+2011.

33 The school was in operation from 1957 to 1974. Brisbane City Council, 'Local Heritage Places: St Ita's School and Presbytery', Brisbane City (2020), https://heritage.brisbane.qld.gov.au/heritage-places/2265.

34 Elaine Martin, 'Social work, the family and women's equality in post-war Australia', *Women's History Review* 12 (2003), 445–68.

35 Martin, 'Social work, the family and women's equality in post-war Australia', 448.

36 For an international overview, see Mary Wyer, Mary Barbercheck, Donna Cookmeyer, Hatice Ozturk and Marta Wayne (eds), *Women, science and technology: A reader in feminist science studies* (New York: Routledge, 2014).

37 Australian Academy of Science, *Women in STEM decadal plan* (Canberra: Australian Academy of Science, 2019).

38 Science in Australian Gender Equity, *Gender Equity in STEMM* (Thousand Oaks, CA: Sage), https://www.sciencegenderequity.org.au/gender-equity-in-stem.

39 Australian Academy of Science (2018), *Putting gender on your agenda: Evaluating the Introduction of Athena SWAN into Australia* (Canberra: Australian Academy of Science, 2018), https://www.sciencegenderequity.org.au/wp-content/uploads/2018/12/SAGE_Report_44pp_SCREEN.pdf.

40 Australian Academy of Science, *Putting gender on your agenda.*

41 Lynn Abrams, 'Heroes of their own life stories: Narrating the female self in the feminist age', *Cultural and Social History*, 16 (2019), 205–24.

42 Research shows that women leaders are predominantly viewed as either competent or likeable, but rarely both. See Catalyst, 'The double-bind dilemma for women in leadership', Catalyst Research, https://www.catalyst.org/research/infographic-the-double-bind-dilemma-for-women-in-leadership.

43 Iain McCalman (2017), 'Linking the local and the global: What today's environmental humanities movement can learn from their predecessor's successful leadership of the 1965–75 war to save the Great Barrier Reef', *Humanities* 6 (2017), 77.

44 Kerrie Foxwell-Norton and Libby Lester (2017), 'Saving the Great Barrier Reef from disaster: Media then and now', *Media, Culture & Society* 39 (2017), 568–81.

45 McCalman, 'Linking the local and the global', 77.

46 Maureen Baker, 'Gendered families, academic work and the "motherhood penalty"', *Women's Studies Journal* 26 (2012), 11–24.

47 John E. N. Veron, *A life underwater* (Ringwood: Penguin, 2017), p. 169.

48 John E. N. Veron, Carden Wallace and Australian Institute of Marine Science, *Scleractinia of eastern Australia. Part V* (Canberra: Australian Institute of Marine Science and Australian National University Press, 1984).

49 Carden Wallace, *Staghorn corals of the world: A revision of the genus* Acropora (Australia: CSIRO Publishing, 1999).

50 Plumwood, *Feminism and the mastery of nature.*

51 Iain McCalman, *The Reef: A passionate history from Cook to climate change* (New York: Scientific American 2013); Celmara Pocock, *Visitor encounters with the Great Barrier Reef: Aesthetics, heritage and the senses* (New York: Routledge, 2020); Ann Elias, *Coral empire: Underwater oceans, colonial tropics, visual modernity* (Durham, NC: Duke University Press, 2019); Braverman, *Coral whisperers: Scientists on the brink*; James Bowen and Margarita Bowen, *The Great Barrier Reef: History, science, heritage* (Melbourne: Cambridge University Press, 2002).

52 Foxwell-Norton and Lester, 'Saving the Great Barrier Reef from disaster, media then and now', 568–81; Kerrie Foxwell-Norton and Claire Konkes, 'The Great Barrier Reef: News media, policy and the politics of protection', *International Communication Gazette* 81 (2019), 211–34; Ally Lankester, Erin Bohensky and Maxine Newlands, 'Media representations of risk: The reporting of dredge spoil disposal in the Great Barrier Reef Marine Park at Abbot Point', *Marine Policy* 60 (2015), 149–61; Libby Lester, *Global trade and mediatised environmental protest: The view from here* (London: Palgrave Macmillan, 2019); Claire Konkes and Kerrie Foxwell-Norton, 'Science communication and mediatised environmental conflict: A cautionary tale', *Public Understanding of Science* 30 (2021), 470–83; Lynne Eagle, Rachel Hay and David R. Low, 'Competing and conflicting messages via online news media: Potential impacts of claims that the Great Barrier Reef is dying', *Ocean and Coastal Management* 158 (2018), 154–63.

53 See Nadine Marshall, William Neil Adger, Claudia Benham, … Lauric Thiault (2019), 'Reef grief: Investigating the relationship between place meanings and place change on the Great Barrier Reef, Australia', *Sustainability Science* 14 (2019), 579–87.

Kerrie Foxwell-Norton is an Associate Professor of Communication and Media at Griffith University, where her work is supported by the Griffith Centre for Social and Cultural Research and the Griffith Climate Action Beacon. Her research expertise is environmental communication, where she explores the politics of community, culture and nature, with a particular focus on coastal and marine environments.

Dr Deb Anderson is a journalist and academic born in Far North Queensland, now based in Melbourne. Her research at Monash University draws from oral history, journalism and ecofeminism to explore the lived experience of extreme weather in an era of politicised knowledge on climate change. She is the author of *Endurance: Australian Stories of Drought* (CSIRO, 2014).

Anne Leitch is a science communication researcher who began her career with jobs that took her all over the Great Barrier Reef counting crown-of-thorns starfish, recording fish behaviour and identifying reef invertebrates. She now spends her time writing about these things and researching climate change adaption and community resilience to coastal change.

The Daintree Blockade: Making (radio) waves

Bill Wilkie
billwilkie2012@gmail.com

Radio log 11/8/84
D5 crossing creek under Timbertop's tree ... continues to fill the creek crossing ... If he continues to fill it high enough the D10 should go through. Looks like a moonscape where the dozers are working.

Timbertop's other alias is Gummy, a well-known rainforest warrior of many campaigns. The radio log gives a live blow-by-blow account of the Daintree Blockade, of police heavy-handedness and council bulldozers destroying the fragile ecosystem. The blockade was started in response to Douglas Shire Council building a road through some of the last tropical lowland rainforest in Australia.

Gummy had fallen in love with the rainforests around northern New South Wales and had travelled to Tasmania for the Franklin Dam protest. He arrived at Cape Tribulation driving a rainbow painted bus with the Nomadic Action Group, all experienced hard-core activists. When the physical blockade at Daintree collapsed on day one, Gummy was one of the first up the trees to commence a tree-sitting protest. He picked a strategic tree right in the path of the road – work would be held up so long as Gummy stayed there. Once Gummy was in position, a two-way radio was hoisted up the tree to link him into the radio network.

The radio log is a record of a radio network that the protesters set up to communicate between outposts along the construction site. The Police and Council could never match the communication systems and it gave the protesters the edge on getting information out to the rest of the world, including the media and politicians. This was one advantage the protesters had, up against the Queensland Police Force under the rule of the Joh Bjelke-Petersen's National Party government, they were consistently out-numbered and out-gunned.

The network was set up by Tiny Toohey, a former farmer from the New South Wales Southern Highlands. Older and more conservative than many of the other protesters, Tiny provided some practical knowhow and a steadying influence. When it came to the radio network, Tiny had enough experience to figure out the logistics of 'bending' the signal around the Cape Tribulation headland and sending it back to Mossman via a relay network including a radio stationed on a yacht.

Three of the small radios survive: 'Contact professional Transceiver' is printed in silver on a shiny blue background. The three radios are the only original items that

Figure 1
Three of the radios used by the Daintree protesters to maintain communication and outwit the Queensland police. Photograph provided by Bill Wilkie.

I have from the Blockade. Much of the rest was pilfered by council workers, destroyed or left to rot in the tropical weather. The Blockade site itself is now disappointing — plaques that are erected to recognise the event are continually vandalised and removed — the most prominent object there is a warning sign for motorists about the condition of 'the four-wheel drive only' track that pushes deep into the rainforest.

The radios themselves are small blue metal boxes, about 18 centimetres long. The collapsible aerial pulls out to 1.5 metres in length. They are generally well preserved, all with a black case with clips and straps all working, though two of the radios are rusted, and I suspect these spent some time up the trees or at the outposts along the track. The best-preserved radio has writing in black marker: the words 'A. I. Toohey, Mowbray No. 3' can be made out along with some other indecipherable writing. I wonder if one of them is the radio Gummy held while perched precariously in his tree, relaying news while bulldozers worked on the forest floor below him.

Radio Log 13/8/84
Timbertop calls for police protection as council workers clearing trees around him. Dozer working very close to his tree and shaking it badly ... Work continuing, dozer bumped tree three times. Police watching on.

The radios came to me from a friend of Tiny's. Tiny died years before I started writing my book but thankfully left his archives to the National Library of Australia. In the archives was a copy of the handwritten radio log of the protest, about 60 pages, much of it recording the actions of the workers and the concerns and observations of Gummy and the other tree-sitters.

Discovering the radio log was vital for my research. It was my way into the story. Using the radio log I could put a protester into the heart of the action, detail the movements of the police and the bulldozers, and put a date to an event that someone had recalled during an interview.

Without the radio log, telling the story of the Daintree Blockade would have been a lot more difficult. The detail, so important in a story like this, would be lacking, and with it the certainty that what I was writing was as close as possible to the actual events. The radios that are in my possession are my connection to my part in the story — the telling of it.

After five days in the tree (an Australian protesting record at the time), Gummy came down; council workers were prepared to divert the road around him, destroying more of the forest. Gummy was arrested, the Blockade was abandoned, the bulldozers pushed on into the rainforest and the road was built.

The protesters did, however, have the last laugh. During the opening of the road, a grand procession of vehicles got stuck between two ranges when it rained and the track became impassable. A hotwired bulldozer was used to get busloads of people out, working well into the night. The following morning, newspapers around the country led with the story: 'Bigwigs bogged ...' read *The Australian*.

Blockade Log 'CLOSED' 23.59 9/10/84

Bill Wilkie's first book, *The Daintree Blockade*, won the Premier's Award at the Queensland Literary Awards in 2017. His writing has featured in *The Saturday Paper*, *Griffith Review*, *QWeekend* and *Ecotone*. His next book is about the Cedar Bay drug raids of the 1970s.

Drawing a line in the sand: Bioengineering as conservation in the face of extinction debt

Josh Wodak
j.wodak@westernsydney.edu.au

Abstract

What conservation could possibly become commensurate with the rates of human-induced biophysical change unfolding at the advent to the Sixth Extinction Event? Any such conservation would require time-critical interventions into both ecosystems and evolution itself, for these interventions would also require domains of risk and ethics that shatter normative understandings of conservation. Yet a line appears to have been drawn in the sand against such experimental conservation. Holding the line will retain conservation practices that are null and void against the extinction debt facing multitudes of species. Crossing the line would invoke scales of bioengineering that appear abhorrent to normative morality. This article explores the question of whether this line in the sand could, and should, be crossed through a detailed case study of current and proposed conservation for endangered *Chelonia mydas* sea turtles on Raine Island, a small coral cay on the Great Barrier Reef in Australia. *Chelonia mydas* and Raine Island are presented as synecdoche for conservation across diverse species across the world because turtles are among the most endangered of all reptiles and Raine Island is the largest and most important rookery in the world for this species. With such lines disappearing under the rising seas, the article contemplates the unthinkable questions that our current situation demands we ask, and perhaps even try to answer.

Game on

Mertle, Tokolou and Turturi are three green sea turtles of the species *Chelonia mydas* (Figure 1). They were named in 2016 following their capture by park rangers and conservation biologists, who glue satellite tags to their shell tops and spray-paint large 'X's across their shell centres. Satellites then monitor their behaviour for around the next twelve months, until shell growth dispels the tag, causing it to fall to the sea floor — now anonymised once more. But this is not their first capture, track

Josh Wodak

Figure 1.
Chelonia mydas, Raine Island Recovery Project, Raine Island, Australia, 16 April 2016.

and tag: more detailed data show that, for Mertle, this goes back to 1992, for Tokolou to 2006 and for Turturi to 1984.[1]

In the more recent taggings and trackings, however, the gameplay and its stakes are altogether different. These three turtles are the only ones named in a publicity campaign for an experimental conservation project in 2016. The relevance of the traditional, familiar forms of conservation — 'yesteryear' conservation — is disappearing at an exponential rate. The conservation ideas and concepts of, say, 2006, 1992 or 1984 are rendered completely redundant because, for these turtles as well as life at large, the world of a few decades ago may as well be another geological and climatological epoch. What, then, is conservation to become, given that this new epoch is manifestly incomparable to the halcyon yesteryears of the twentieth century? And where do we draw a line in the sand if conservation in the Anthropocene amounts to bioengineering, not only in ecosystems but in planetary-scale evolution itself?

These three turtles have borne, and still bear witness to, this transition from recent conservation ideas to current conservation ideas, since what is at stake is not the existential plight of their sole species but rather the fate of their entire superfamily *Chelonioidea*. As a team of conservation biologists remarked in their 2018 article 'Where Have All the Turtles Gone, and Why Does It Matter?':

> The fate of turtles is especially tragic in light of their distinction as paragons of evolutionary success. They survived everything nature could throw at them from both earth and outer space (for example, the asteroid that wiped out the dinosaurs), but will they survive modern humans?[2]

Looking at conservation ethics through a deep-time lens, these three turtles are synecdoche for the plight of life-at-large at the advent of the Anthropocene. Their "fate" entwines their superfamily's 120 million-year record of enduring not only the two mass extinction events prior to this one currently unfolding, but also all the many varied unnamed ruptures that are part and parcel of long-term survival. Yet their "fate" seems to hinge upon radical and risky experiments to construe a conservation that could counteract their extinction debt.

The response to the plight of Mertle, Tokolou and Turturi sought to understand the question of 'Where have all the turtles gone, and why does it matter?' by tracking where turtles are going at present. Peering down on them from satellites over the period 2016–17 confirmed some staggering feats of green sea turtle migration, including that they swim up to 2,500 kilometres to reach their nesting site, from as far away as Vanuatu, Indonesia and Papua New Guinea. Each nesting migration is an unvaried return to the exact beach of birth, navigating via Earth's magnetic field — a field that has periodically flipped polarity many times, for when one's habitation extends over 120 million years one's homing beacon fixates on points in space and time as fluid as the rise and fall of entire continents.

For these three turtles, their sole destination is a drop in the ocean: Raine Island.

Raine Island vs artificial rain

Raine Island is a microcosm for how life did not happen on, but rather to, Earth. After all, the island is biogenic. Like all coral cays, its origins lie not only in a lifeform (coral), but also in the death of this same lifeform, composed of skeletal coral remains. Once Raine became an established land mass not only year-round, but year-in, year-out, life then happened to this microcosmic Earth on a whole other level. For coral is not the only way life (or at least life's excrement) is literally woven into the island's fabric itself. Next came an Earth suffused with life from the top down: shit happens, and it happens from the sky.

Biogenic contributions to creating Raine came from guano, created when dried bird shit accumulates on a semi-solid foundation — rocks or sand above sea level. Guano reacts with existing sand, sediment, and water to make top-down bedrock, a mirror process to how coral aggregates sediment to make bottom-up bedrock. The guano bedrock then lays the foundation for soil and grass to form, which adds to existing attractions for birds to nest. This creates more guano, which amplifies the cycle of shit-fuelled 'land ho!'

To the naked eye, it appears obvious how corals change cay geomorphology: no coral means no island. But the way guano can change cay geomorphology is hidden from plain sight. Chemical reactions started by the bird shit catalyse 'a unique form of reef island in which a phosphatic cap formed from the downward leaching of guano plays an important part'.[3] While this process is endemic to all phosphatic cap islands, "unique" refers only to Raine for reasons borne out by this island's creation and features, which make it a microcosm of Earth.

The sentence comes from *Raine Island: Its Past and Present Status and Future Implications of Climate Change*, a 2008 report by geomorphologist Professor David Hopley. The sentence's abrupt ending begs a cascading series of questions: 'An important part' in what? Important in catalysing Raine? Important for Raine's resistance to oceanic erosion? Important for marine and terrestrial life, which depend on Raine for breeding and nesting?

The possible answer — a definitive answer is absent from the 101 pages of the report — encapsulates why the guano-phosphate cap has life-and-death consequences for turtle nesting. Given turtles' drive to nest only where they are born, nesting sites are the existential thread connecting successive generations. And given that Raine is the largest remaining *Chelonia mydas* nesting site in the world, and thus the most important green sea turtle rookery, the island makes for a telling microcosm of the stakes at play in conservation at large.

Since geographically distinct populations of *Chelonia mydas* do not interbreed, each is effectively constrained to its respective nesting site. The only other population in Australia, some 2,000 kilometres south in the Southern Great Barrier Reef, makes Raine the primary preserve of the entire Northern Great Barrier Reef population. In addition, Raine's rookery has the longest known use in the world, having nurtured turtles for at least the last millennium.

Thus the story of Raine Island takes us from Mertle, Tokolou and Turturi to *Chelonia mydas,* then to turtles at large, and beyond to life at large, because turtles are only one of many life forms dependent on this drop in the ocean. Some 130 years of European observations have identified 84 bird species nesting on Raine, making it 'one of, if not the most important tropical seabird nesting site on the Great Barrier Reef'."[4] And, like green sea turtles' mammoth migrations, the birds connect Raine to distant corners of the globe. This island, a mere 3 square kilometres, 100 kilometres off the mainland, is globally interconnected to distant corners of the planet by the migratory feats of those nesting there. If Raine is a drop in the ocean, it shows how all drops are connected far and wide.

Mertle, Tokolou and Turturi were born before 1981, when the global importance of the refugia was first recognised by establishing the Raine Island Corporation. These three turtles herald the last generation before the geomorphology of the phosphatic cap began to play 'an important part' in the survival of their species. By the time Hopley published his 2008 report, the stakes around the refugia's protective status had been raised, and the island's management had been transferred from the Corporation to the Queensland state government just one year earlier. The stakes were raised again when it was subsequently reclassified as a 'Nature Refuge' and then as a 'National Park (Scientific)'. The latter holds the most stringent status and highest level of protection in Australian law, including limiting all access to 'scientific research and essential management only'.[5]

Talk, though, is cheap. While it is already dubious whether such strictures hold up on the scale of Raine, on a planetary scale all legal protections are revealed to be null and void for the unfolding rupture of life on Earth. Arbitrary boundaries that demarcate a 'Nature Refuge' are empty gestures against underwater heatwaves-cum-heatfloods, when warming waters never subside and rising waters never recede.

Ramping up

If Raine is a microcosm for the way life happened *to*, rather than *on*, Earth, turtles demonstrate just how entangled biotic forces are with the abiotic planet itself. Such entanglement is especially problematic for fledgling turtles, for whom Earth is both incubator and protector. Burrows keep eggs warm enough, cool enough and dry enough, and hold them deep enough and long enough to hatch without being cooked, drowned, eaten or prematurely unearthed. Each existential challenge is just a smattering from biophysical limits to life. Too hot means dying inside the egg. Too cold means there is insufficient growth inside the egg. Too much water means drowning inside the egg. Too little water means dehydrating inside the egg.

Such is life (and death) for green sea turtles. But on Raine the 'unique form' of solid phosphatic rock poses distinct existential challenges. During nesting season, upwards of 20,000 individuals may burrow at the same time, redistributing so much sand over the island that they unintentionally construct ramps from the beach to the raised phosphate centre. In search of better sand for burrowing, or due to overcrowding of other turtles on the beach, some climb these ramps to nest on the raised central platform.

These ramps are the only means up or down, because the edge of the phosphate platform has eroded, creating a cliff with a metre or so drop to the beach below. Platform hatchlings then need the accidental ramps created by adults to still be there so they can get down onto the beach and into the sea. The first existential challenge for those platform birthing or hatching is to be able to ramp down, with all the entailed vicissitudes, given that a ramp constructed today is ephemeral, given the winds, hurricanes, surf swell, or burrowing by subsequent turtles over the 60 days of incubation.

For adults, there is an additional predicament, both cognitive and proprioceptive. *Chelonia mydas* live their whole lives in water, other than when on Raine. Being on land is the exception to their rule: following birth, the first 40 years are spent at sea. Having reached maturity, females then return to the beach of their own birth once every two to six years for a nesting season, nesting four to six times per season. Having swum up to 2,500 kilometres to return to Raine, they haul a 130-kilogramme torso across tropical sands in 35–40°C heat, negotiate their way through the thousands of their kind doing likewise, locate a site, then burrow for hours on end to create a nest. By the time they finish laying their eggs, their energy levels have plummeted, making for a dehydrated and delirious return to sea.

With energy repositories barely sufficient to provide motion toward the sea, the wherewithal to search for a ramp is often found wanting. Unable to negotiate the cliff edge, many exhausted turtles fall over, landing on their back on the hot beach sand. Back-to-front turtle flipping has subsequently become an endeavour whenever scientists and park rangers visit Raine during nesting season. They walk the cliff base perimeter, stepping around those nesting on the beach, to manually flip hundreds of turtles back onto their fronts. At first, staff drag them on surfboard-like platforms across the beach to get them back in the sea. As demand increases, they introduce little diesel-powered tractors to wheel them back into the sea.

Manual turtle flipping is a reactive intervention, meaning the reaction follows after an event has occurred. Reactive interventions suffer from bleedingly oblivious intrinsic limits, especially when applied to human-caused environmental damage. First we create catastrophes, then we reactively attempt clean-ups that are often completely insufficient and haphazard. A reactive mentality is already problematic for discrete incidents with proximal cause-and-effect chains, such as oil spills. But it is absolutely delusional when confronted with distal non-linear phenomena that are discrete in neither space nor time, such as climate change. Coaxing oil back into barrels is self-evidently hazardous. Coaxing greenhouse gases back into earthly sequestration is self-evidently preposterous.

The logical alternative to reactive interventions is also fairly self-evident. Proactive interventions attempt to anticipate, rather than react. First extrapolate from the present to an anticipated future consequence of present phenomena, then intervene to thwart the undesirable future from eventuating. Or so the logic goes, where the proof in the pudding is that the originally anticipated future does not eventuate. The premise is that pre-intervention is better than cure. On Raine, the first such proactive intervention can be seen in the low fences constructed around the cliff edge to stop turtles going over. Now a turtle is prevented from flipping onto its back, provided the fence exerts sufficient resistance.

Just as turtle flipping and fence construction are surface-level interventions on this earthly microcosm, there is much more than meets the eye in terms of what conservation could, or should, become if it is to remain relevant to the velocity of biophysical change afoot. In particular, there is the inheritance of an historical legacy by which our species acquired geological agency, and by which the ethics of current conservation debates ought to take heed.

Staging an intervention

In the case of Raine Island, the legacy that prefigures current conservation debates dates from 1842, when a barque hailing from England was wrecked on a reef 40 kilometres from Raine.[6] Many such ships ran aground on the Great Barrier Reef, which led the British colonial authorities to build a beacon on Raine, in the form of a non-light-emitting lighthouse. The island was selected not so much for its position, but rather for its phosphate cap, which guaranteed its relative year-to-year landmass stability and also provided on-site building resources. In 1844, an expedition began building the beacon with rocks mined from the phosphate cap and timber salvaged from the 1842 wreck.

So followed a settlement of convicts consigned to build the beacon, and their captors who enforced their labour. The island peak was now the tower, neighboured by an inverse indention from the quarried central phosphate platform. Following the coral and bird shit biogenesis, Raine's first non-indigenous incursion constituted a distinctly modern form of anthro-genesis.

The legacy of this anthro-genesis ran amok for the following few decades, once he beacon was built and Raine was once again uninhabited and unvisited. When intensive guano mining began in 1882, tens of thousands of tonnes of guano were mined from the island's central platform to manufacture phosphorous that was shipped all around the globe. Raine's shit-catalysed topography was distributed onto farming fields far and wide, as phosphorous was foundational to developing industrial agriculture. Unlike the explicitly coerced beacon builders, coercion was now implicit between Chinese labourers and their European overlords. The distinction alludes to the vastly differentiated agency, will and responsibility for how the anthropos acquired the Anthropocene's geological agency.

The legacy of this second stage of anthro-genesis on Raine ran amok for the following few decades, as Raine became uninhabited once more after all mining had been exhausted. In combination, both stages of anthro-genesis constituted a biogeochemical upheaval of Raine. With the upper guano removed, rainwater could now penetrate the raised central platform substrate. Chemically reacting with the phosphate, seeping rainwater then eroded the substrate away, in turn eroding the cliff edges of the central platform. Further, the bryme bedrock that lies directly under the beach sands where turtles nest has progressively shifted under their burrowing area, by way of phosphate transported by water that has leached through the eroded cap, effectively making a solid platform under the beach sands.

The shifting bedrock appears to be the terminus for David Hopley's 2008 statement that 'the downward leaching of guano plays an important part'.[7] Conservation biologists were initially baffled about the plummeting hatchling numbers, hence Hopley's ambiguity about what was the 'important part'. Subsequent research has revealed that since the bedrock is not porous, the water table is increasingly rising up the beach area, where turtles build their nests. The bedrock thus retains falling rain under the beach, and it exacerbates sea water rising through the beach sand. Thus, from coral–bird shit–water–sand–human–industrial intermingling, a rising sea water level was drowning the vast majority of incubating eggs.

By 2014, the ratio of turtles hatching versus drowning in their eggs had become critical. Whereas 370,000 hatchlings represented an average nesting season, all other things being equal, conservation biologists estimated 2,500 hatchlings that season.[8] Government agencies reactively intervened by experimentally increasing egg distance from the water table. Transporting massive engineering vehicles onto Raine, they raised a 100 x 100 metre beach section 1 metre, redistributing the equivalent of six Olympic swimming pools of sand from the front beach area to the rear. With the experiment indicating a less dire sink-to-swim ratio, the state government proposed a larger scale beach reprofiling.

Instead of fronting up the A$8 million required to do so, however, the government advertised for private sponsorship. David Attenborough lent his voice to this fundraising video, inviting the viewer to 'be part of the largest green turtle recovery project in history'.[9] Attenborough has a long association with Raine, having first filmed there in 1957 and, like a turtle returning to its birth site for its first nesting season, he returned to Raine a half-century later, filming the 2014 fence-building and beach-raising interventions. Featuring the experiments in his major three-part BBC series *Great Barrier Reef*, the projects suddenly received international attention. A consortium of companies, NGOs, state agencies, universities and Traditional Owners responded to the call for desperate measures, and with funds secured, the Raine Island Recovery Project launched in 2016.

The title is unwittingly ironic: recovering Raine by literally re-covering the beach in its own sand. A larger re-covering intervention in 2017 bore a title befitting the neoliberal conservation: Operation Sand Dune. After all, such conservation is *Nature Inc: Environmental Conservation in the Neoliberal Age*, as the authors Bram Büscher, Wolfram Dressler and Robert Fletcher lament.[10]

Mertle, Turturi and Tokolou were tagged and tracked as part of analysing the comings and goings of adult turtles during the Raine Island Recovery Project. Indigenous schoolchildren named Mertle and Tokolou from their respective local indigenous languages. But it was multinational fossil fuel company BHP that named Turturi, a fringe privilege gained from the company's principal funding of the whole five-year project. This is conservation through bioengineering, outsourced by a state government, funded by a fossil fuel company. A deal with the devil it may be, but

Figure 2.
Chelonia mydas, Raine Island Recovery Project, Raine Island, Australia, 8 August 2016.[15]

conservationists who have wagered this game for decades are putting all options on the table because the floor is already soaked through with the rising seas.

One could take the moral high ground and protest the bitter irony of the funding source. But the moral high ground is not as obvious as refusing to stand on the re-profiled beach area. Without that sponsorship, the second beach level raising and ancillary experiment support would not have occurred in time. When populations plummet, so does the luxury of time in which to respond. The question posed to conservation, then, concerns not just where we draw a line in the sand, but when.

Exclamation. Point. Extreme.

When Attenborough returned to Raine in 2014, he drew a perplexing line in the sand, the demarcation of which made a mockery of any protests against intervention, while also revealing that any such interventions amount to empty gestures when viewed on any meaningful scale.

In his 2014 documentary, Attenborough unintentionally shows up human hubris as a hollow conceit when he profiles the work of conservation biologists who count hatchlings from the re-profiled beach area that they have steered into a shallow trench. Members of the team from the Great Barrier Reef Marine Park Authority lift each hatchling from the trench, putting them onto the beach to continue their shell-to-sea journey. Attenborough concludes the segment about Raine by interviewing the project leader, Andy Dunstan, on the beach-raising experiment:

> It's confirmation that Andy and his team have found the right way to restore this vital breeding area. But for the young hatchlings, the trials of life have only just begun. Each new arrival will have to make a perilous dash to reach the ocean. Now they're on their own. Andy and his team must not interfere at this stage. Inevitably, the tiny, defenceless hatchlings attract scores of predators.

This statement draws a line in the sand, on the basis that 'Andy and his team must not interfere' once the hatchlings leave the trench. It appears as if the tiny hatchlings have crossed some invisible line, leaving the protection of the human hands that have raised their nesting beach so they did not drown as eggs, flipped their parents onto their backs so they did not die of dehydration, and built fences and even trenches to aid adults and hatchlings respectively. But now the hatchlings return back to the yonder side of nature, the one that is 'red in tooth and claw',[11] with all human aid withdrawn.

For every 1000 hatchlings, only 70 make it to the sea. On the journey from shell to sea, crabs or birds eat those whose 'trials of life' end after they 'have only just begun'. For every 1000 that make it to the sea, only one makes it to adulthood. On the journey into the sea, tiger sharks wait so close to the shoreline that their entire backs are exposed above water. Granting safe passage from shell to sea would thus go a long way towards greater numbers reaching adulthood. If aiming to help plummeting populations, why go only so far as flipping individuals back onto their front and propping up portions of their nesting area over the rising seas? Socially okay to capture, tag, and track. To manually flip. To transport on tractors. To install clifftop barriers. To raise the beach height. But, having partially restored shelter to their refugia through all this intervention, 'now they're on their own. Andy and his team must not interfere at this stage.' Why is everything so far considered acceptable, but sheltering hatchlings from predators on their shell-to-sea journey is not?

Worse still, why tell ourselves we made right when so much more remains to even remotely get things back up the right way round? What prompted Attenborough to set the limit to intervening here? It is not as if the Raine Island Recovery Project has any means to redress the myriad existential challenges facing coral beneath the island, turtles within the island and seabirds atop the island. Moreover, the acuteness of the situation is well illustrated by the fact that the conservation mentality driving the project shows the same insatiable urge to cannibalise Earth as the industrial capitalism (here in the form of a fossil-fuel company) that it depends on for its primary funding. Hopley demonstrates this in his report's sobering rationale for Raine's present status and future implications of climate change:

Nearby cays … before they disappear their value may be in providing the sand for any replenishment on Raine Island. This will be a no loss position, as within a few years as sea level rises, these cays will disappear completely and their sand resource lost forever.[12]

Thus far, the universities and companies who paid to attach their names to such conservation-as-bioengineering have celebrated so-called success in their marketing and public relations. Yet the fact that the neighbouring cays are disappearing under the water remains set aside. Even worse, they remain entirely mute on the acute existential threat that was subsequently unearthed when the project was only halfway through its tenure.

Like reptiles in general, turtle sex is determined by egg incubation temperature. Below a certain sand temperature threshold, turtles are born male; above it, they are born female. As adults, Mertle, Tokolou, and Turturi hark from the 1970s and 1980s, when one male *Chelonia mydas* was born for every 6.6 females. By the time the beaches were being raised, this had changed to one male for every 116 females. In response to these findings from her research, turtle scientist Camryn Allen declared in 2018 that 'this is extreme — like capital letters extreme, exclamation point extreme … we're talking a handful of males to hundreds and hundreds of females. We were shocked.'[13] From this point on, all talk, though still cheap, should be in capital letters all the same. EXCLAMATION. POINT. EXTREME.

When Attenborough crouched over hatchlings born in the re-covered heightened sands, he declared that, 'for the young hatchlings, the trials of life have only just begun'. Beyond the usual proximal predators, their trials of life are actually at the behest of the distal suspect that is murder most foul: anthropogenic climate change. Behind the far more pernicious threat to the turtles' ongoing existence lies the same non-linear causation that is raising the water table up the beach. From climate change: too wet, they drown; too hot, they cook.

Having drawn a line in the sand about not interfering in hatchlings' shell-to-sea passage, how would conservation ethics prefigure any intervention that could flip their male-to-female sex ratio back towards 1:100, let alone 1:50 or 1:10? It is here, in bioengineering as conservation, that the stakes find their fullest expression in risk, complexity and uncertainty: intervening directly into evolution itself to forestall extinction. Such interventions are about catalysing effectively instantaneous phenotypic and/or genotypic modification for future descendants to inherit the modified traits. Say a discernible phenotypic or genotypic trait takes ten millennia to appear across a species, and scientists induce a discernible inhered trait in ten years in a population of the species. From an evolutionary time perspective, that is effectively instantaneous.

Therein, redressing the sex ratio would necessitate radical interventions into both ecosystems and evolution. The risks, complexity and uncertainty in this predicament are palpable. The timeframe to act is now — or never. The consequences quite literally determine the procreative viability of sea turtles, due to the extinction prospects arising from one male for every ten females, or one for every 100, or one for every 1000 … Against the rapidly diminishing timeframe for any efficacious intervention, what would such bioengineering entail? More to the point, assuming any such means, vexing questions arise as to whether we could or should. The burning question thus becomes: What would such risky and radical conservation look like for Raine's turtles?

End game: Buck. Stop. Here.

Once again, even for matters such as the utter urgency of reconceiving conservation-at-large, an historical legacy remains to be heeded, like Raine's beacon. Because the climate system has roughly a five-decade lag between emissions and their consequences manifesting, the change in the Raine Island turtles' sex ratio from 1:6.6 to 1:116 in 2018 was due to emissions made around the time Attenborough first visited Raine in 1957. So *now* is already entirely too late.

This means that to be anything other than an empty gesture, only a seismic (and successful) intervention in the *Chelonia mydas* genotype could stop the sex ratio from going from its current 1:116 to 1:1160, all other things being equal. Let alone bring it back from 1:116 to 1:6.6. Will Attenborough lend his voice to campaigns to sculpt the *Chelonia mydas* genotype via millions of dollars of corporate donations? Will there be public outcry or even public debate about such proposals?

Instead, our *now* is stuck in the meantime of ad hoc reactive interventions. One is to relocate eggs to cooler sands, including on neighbouring islands. Another is to lower sand temperature by modifying the albedo ratio — say, by adding light-coloured sand to nest tops, sand that could not stay on top for long, given the how much sand the burrowing turtles redistribute. Uncertainty also abounds regarding whether imported sand would bring fresh existential challenges to incubation mortality on the microbial scale, from micro-organisms that inhabit coral cay sand.

To circumvent these ground-level problems, another proposal simply moves higher up, suggesting shade cloths be used to modify the albedo ratio, suspended above the vicissitudes of sand morphology and terrestrial microbes. Such experiments have already taken place in a handful of key rookeries across the globe, including over the Mon Repos rookery in the southern Great Barrier Reef, in an initiative of the Queensland Department of Environment.

While shade clothes demonstrably result in proportionally more males being born, Raine's logistics are unwieldy. Consider 20,000 turtles arriving and departing all hours of the day, weeks on end over nesting season, which is also hurricane season. Having the right nest in the right area at the right time becomes a game of cat and mouse — unless human intervention extends to completely covering these so-called natural environments in shade cloth tents. While this is under consideration for Raine, due to the remote location and unwieldy topography, such structures would turn it into a triage centre, requiring round-the-clock vigilance on par with a hospital emergency ward.

To circumvent these immediately-above-ground-level problems, some suggest going further upwards, above the vicissitudes of shade cloths and endemic hurricanes. Precipitating artificial rain by cloud seeding is nothing new, although it prefigures planetary-scale proposals to engineer the climate, to avoid runaway climate change. Will BHP or another fossil fuel company sponsor such intervention? Does this not show up the ludicrous mindset of funding technofixes via the same industrial society that catalysed the rupture to begin with?

In any event, both shade cloth and artificial rain are still only ephemeral interventions. While they are at the forefront of existing conservation experiments,

Josh Wodak

they do not remotely amount to the seismic *now* that can avert the impending *never*. As regards this most vital rookery-turned-triaged-emergency-ward, ephemeral interventions are band-aids to the insoluble condition known as evolution. Given how precipitous the sex ratio already is, further proposals to systemically redress this sex ratio are far more insidious.

In relation to turtle conservation, synthetic biology would entail discovering the genes that assign sex according to temperature, modifying them, then using gene drives to get the modification inherited across generations. For example, sand that is 29°C produces more females than males. At 33°C, all hatchlings are female. At 23°C, all hatchlings are male. Synthetic biology would modify the genome so that the eggs of future descendants would become male at a higher temperature, say 31°C.

To date, no such research has been publicly proposed. Recall Attenborough drawing a line in the sand about stopping the birds and crabs from eating hatchlings on their journey from shell to sea. How would he, or younger generations of conservationists or environmentalists, draw a line in the sand about such conservation? In the currently unfolding rupture, such conservation can neither be reduced to a mere fantasy, nor can its prospects for yielding anything efficacious be considered even remotely real, or commensurate with the timeframes available.

A much larger historical legacy looms here too, one that concerns the complete tenure of turtles themselves. Throughout their existence, turtles have accommodated the comings and goings of things essential to their livelihood, such as nesting sites, the seagrass species they eat, currents and ocean temperatures. Their adaptive capacity to avoid maximum sand temperatures includes nesting at cooler times of the year and changing to beaches with a lower solar albedo. The same being that is so insistent on returning to nest at its birthing beach, even when it proves perilous, is also able to collectively revise that instinct and switch to new times of the year and new locations when the timeframe is sufficient to accommodate the gradual pace of such evolutionary change.

Ironically, the turtles' instinct to return to their exact birthplace makes them less able to accommodate durational change, such as the comings and goings of entire rookeries. The antiquity of their tenure on Earth self-evidently attests to their ability to shift *en masse* to different islands, though their instinctive return to birth sites also attests to their propensity to nest in sites that have since become perilous. A rookery that disappears within a century may not provide enough time for turtles to shift *en masse* to a new breeding ground. But disappearance over a millennium may suffice. As always, adaptation is a matter of time.

The same principle applies to a food source disappearing, or ocean temperature substantially changing. Yet seismic and catastrophic change of the stochastic and non-linear variety is nothing but normal. A 2018 hurricane eviscerated an entire Hawaiian island that was a sea turtle refugia — even if we set aside the inarguably just, but at this point ineffective game of poring through fragments to determine the extent of human-induced amplification of that hurricane, shift happens just as shit happens.

Yet the timeframe for playing dicey conservation games bears next to no relation to either saltation moments of evolutionary exuberance — such as when turtles adapted to prior rapid biophysical change — or atmospheric turbulence that could eviscerate Raine in a single hurricane season. The game can, in fact, be placed squarely in the timeframe between the rise of nineteenth-century colonial and industrial activity and today. The seismic local impacts of the 1842 beacon construction and 1880's mining took 180 years to ensue. They offer scaled-down proxies for the cataclysmic planetary impacts stemming from the five-decade lag between emissions and their demonstrable effects on the climate system.

Caught as we are between historical emissions and contemporary attempts to deal with their aftermath, we can only foretell that greater volatility and vulnerability for respective abiotic and biotic bodies lie ahead. Just as the legacy of nineteenth-century human interventions into Raine have been made manifestly apparent in the island's geomorphology, now there is no doubt about the anthropogenic legacy that foreshadows the future.

For Raine, such a spectre will come home to roost much later than any of its neighbours. Once climate change drowns all the surrounding islands, Hopley predicts that 'Raine Island is likely to be the last cay to disappear in the northern Great Barrier Reef'.[14] This depends upon the unique phosphate cap continuing to provide a bulwark against erosive ocean currents, and coral continuing to replenish aggregating sediment, which in turn means Raine is ultimately dependent on the ongoing living and dying of coral.

An end to living coral means an end to replenishing the dead coral that provide aggregating sediment piled into the raised area that makes Raine's seashore. Raine's story therefore begins and ends with a rock, albeit a living rock of abiotic progeny: coral corpses.

But Raine's other story — the human one — begins and ends with another, different kind of rock: the beacon. That piled high aggregate of mined phosphate cap now stands at a maximum of 7 metres above sea level. One day, when the rest of Raine has gone under, a beacon poking up out of a flooded sea is all that will be exposed, its sociopathic means of creation thus obscured, like the myths by which the civilisation of its day lived and died.

Acknowledgements

This research was funded by the Australian Research Council Centre of Excellence in Synthetic Biology (CE200100029).

Notes

1 Andrew Dunstan and Katharine Robertson, *Raine Island Recovery Project: 2016–17 Season Technical Report to the Raine Island Scientific Advisory Committee and Raine Island Reference Group* (Brisbane: Queensland Government, 2017).

2 Jeffrey Lovich, Joshua Ennen, Mickey Agha and Whitfield Gibbons, 'Where have all the turtles gone, and why does it matter?', *BioScience* 68(10) (2018), 772.

3 David Hopley, *Raine Island: Its past and present status and future implications of climate change: Project report* (Townsville: School of Earth and Environmental Sciences, James Cook University, 2008), p. 1.

4 Hopley, *Raine Island*, p. 13.

5 Peter Beattie and Lindy Nelson-Carr, 'World's largest green turtle rookery given highest protection status', ministerial statement, Record of Proceedings, First Session of the Fifty-Second Parliament of Queensland, 22 August 2007, p. 2727.

6 Hopley, *Raine Island*, 68.

7 Hopley, *Raine Island*, 1.

8 Andrew Dunstan, quoted in Neil Mattocks, 'Natural history and research and management of Raine Island's green turtle rookery', eAtlas, https://eatlas.org.au/ts/raine-turtles.

9 David Attenborough, quoted in the 'Be part of the largest green turtle recovery project in history' poster, Department of Environment and Heritage Workshops, Brisbane, 2016.

10 Bram Büscher, Wolfram Dressler and Robert Fletcher (eds), *Nature Inc.: Environmental conservation in the neoliberal age* (Tucson, AZ: University of Arizona Press, 2014).

11 Alfred Lord Tennyson, *In Memoriam A. H. H.* (London: Edward Moxon, 1850).

12 Hopley, *Raine Island*, 48.

13 Camryn Allen, quoted in Craig Welch, '99% of Australian green sea turtles studied turning female from climate change', *National Geographic*, 8 January 2018, https://www.nationalgeographic.com/science/article/australia-green-sea-turtles-turning-female-climate-change-raine-island-sex-temperature; Michael Jensen, Camryn Allen, Tomoharu Eguchi, … Peter Dutton, 'Environmental warming and feminization of one of the largest sea turtle populations in the world', *Current Biology* 28(1) (2018), 154–9.

14 Hopley, *Raine Island*, 48.

15 'Green turtles die on Raine Island, unknown provenance, courtesy Qld Govt', eAtlas, https://eatlas.org.au/media/977.

Dr Josh Wodak works at the intersection of the Environmental Humanities and Science and Technology Studies. His research addresses the socio-cultural dimensions of the climate crisis and the Anthropocene, with a focus on the ethics and efficacy of conservation through technoscience, including synthetic biology, assisted evolution and climate engineering. He is currently a Senior Research Fellow at the Institute for Culture and Society, Western Sydney University, a Chief Investigator at the ARC Centre for Excellence in Synthetic Biology and an Adjunct Senior Lecturer, School of Biological, Earth, and Environmental Sciences, University of New South Wales.

'Tourist fiction': Cassowaries in Mission Beach

Leonard Andy
Djiru Traditional Owner (as told to Dr Valerie Boll)
valerieboll_27@hotmail.com

My name is Leonard Andy and I'm a Djiru Traditional Owner of the Mission Beach area.

Where I live today and where my Ancestors have lived is not the same place. Today the Mission Beach area has become a tourism destination and it has changed the people, our culture. Presently, there are twelve Traditional Owners living in the area, off these twelve, five are still at school.

The cassowary is an important animal to our people and we would like it to be here in the future for our children and children's children to see.

When it comes to planning in conservation and preservation of the cassowary, it seems rainforest people are often an afterthought. We, the Traditional Owners who live here, are getting worried about what will happen in the future in the area with *Gunduy*, the cassowary. Because of tourism development and more people coming to our area, we are losing habitat for the cassowary.

For example, a lot of subdivisions in our area were previously swamps in the 1980s and 1990s; they were cleared and drained and landfill was used to make everything level. These subdivisions were and are still approved by the Cassowary Coast Regional Council. They continue to be the source of a long-term loss of habitat. Also, I don't know of any subdivisions where there are restrictions on pets, such as dogs, which injure and kill cassowaries. Another example is that we've seen changes — for example, where the cassowary has stopped to eat bush tucker from the forest. In some of the built-up areas, the cassowaries come in from the forest to eat bananas, grapes and pears, because these are given to him by new residents that have moved into our area. It seems that people are just copying what has been done by people who came 20 or 30 years before them and the new locals seem to think this is what you do. We know that National Parks have certain rules and fines for feeding cassowaries but we haven't heard any in our area where they've been issued.

Our national parks have laws but not really laws, just policies and procedures, created by the government to implement at the grass-roots level. Some of us, Traditional Owners in the Wet Tropics region, not just my area around Mission Beach, have concerns that the cassowary recovery plan doesn't do much for us, especially when this plan has no real understanding of our culture and spiritual connections to this animal. We understand that science can do a lot of things, but

Figure 1.
Leonard Andy with the 'Gunduy Midja' (cassowary dwelling), 2019, art installation in collaboration with Nina Dawson, Cassowary Festival, Mission Beach. Photography by Valerie Boll.

Figure 2.
Leonard Andy, 'New Wuju' (new bush tucker), 2021, acrylic on canvas. Courtesy of the artist. Photography by Valerie Boll.

now some of us are wary of scientists and science. When it comes to some of our cultural values, we are told by conservationists and scientists that the global values of the World Heritage listing are important to everyone in the world, protecting the cassowary, the forest and the reef for everyone. Some of us, who live here and have been here for a very long time, don't really believe it. With such protection comes economics. Some of us are wary of the tourism operators saying that they are here to protect the reef for everyone. I think they are here protecting their own financial business. For example, we have an 'open season' and a 'closed season' for certain fish, like for the barramundi. Maybe we should have an 'open' and 'closed' season for tourism on the reef instead of being open twelve months a year.

That's why how we feel about the cassowary has nothing to do with tourism because our connection, spiritually and culturally, was there before tourism, before World Heritage, before national parks. Our connection is still here, but I would say it is suppressed under legislation and governance.

Another confusing aspect of managing our land and the cassowary is that we have native title in our area and we have native title rights that are federal, yet some of these rights are only on paper. Also in Queensland there are state rights and it seems as if, in some areas of conservation, it is the state that implements a lot of the policies and procedures, whether it is state or federal on the ground. In practice, you are told one thing by the federal government but then you are told something different by the state government.

We are also told everything is based on the best science available. Some of us question whether or not it is and how we know it is the best science available.

We work with National Parks and the State of Queensland talks to us about co-management of the national parks and also our cultural heritage, which the cassowary is part of. This co-management seems to be at a lower level where we have limited input into policies and procedures. For example, we have Aboriginal rangers who get trained in all the skills they need, but still they are Aboriginal rangers carrying out non-Aboriginal policies and procedures on our land, plants, animals, water. I question whether we have any input working with scientists, because it seems as though with the departments, we get told repeatedly, 'This is based on the best science available.' My problem is that by putting the environment and science together, they are not treated as equal. The science seems to be the ruling factor. We are unable to get any grant money to do things on country environmentally today unless we have supporting evidence based on 'the best science available'. Science dictates what you do or if you do anything at all, not because the environment is in danger or currently being harmed. Our efforts to help are constrained because on the form we haven't 'ticked the box' indicating which scientist was brought in, or is your 'pet scientist'. Furthermore, it also seems like it doesn't matter if you are the Traditional Owners or you have Elders telling stories about their experiences. It doesn't count because it is what the scientist says that matters. People, politicians, heads of departments, these are what they rely on and will listen to. You are listening to one side of the story because we, the Traditional Owners, have another side of the story.

Our conclusion is that we were the first cull, followed by the cull of the environment, the crocodiles. Our agenda is set for us by tourism, but it shouldn't control our lives.

Leonard Andy is a Djiru Traditional Owner currently living at Mission Beach on his traditional land. He creates a number of unique art pieces including paintings, sculptures and wooden artefacts. His attention to detail is evident in the intricate designs painted on his carved wooden swords, boomerangs, spear throwers and canvases.

Epilogue: A reflection on the role of tourism within vulnerable biodiverse reef and rainforest regions – a case-study from Mission Beach and the Cassowary Coast

Iain McCalman
Iain.McCalman@acu.edu.au

It is heartening to see that so many of the scholarly and personal contributions of our special issue should have addressed the complex collisions between culture and nature manifested today within Queensland's Great Barrier Reef and Wet Tropics Rainforest World Heritage Areas.

This special issue has also illuminated for me some specific challenges that confront local communities with whom I've worked in the Mission Beach and Cassowary Coast region over the past decade. Here, halfway between Townsville and Cairns, where 'the rainforest meets the reef', exists one of the most ravishingly beautiful and densely biodiverse yet fragile regions in Australia. Mission Beach and other Cassowary Coast towns have in recent years endured a succession of severe tropical cyclones. Reeling from the destruction of Cyclone Larry in 2006, communities and environments were struck again by Cyclone Yasi in 2011. Between them, these two cyclones damaged houses and businesses, flattened the region's famous Dunk Island resort, smashed both fringing and offshore coral reefs, ripped up acres of rainforest and killed around a third of Australia's largest remaining population of Southern Cassowaries. On top of this, many of the region's surviving local reefs have since experienced climate change-induced coral bleaching. Finally, the COVID-19 pandemic hollowed out remaining forms of tourism here, as elsewhere in Queensland. This convergence of crises has contributed to a major slump within the tourist-based businesses on which the region's economy largely depends.

At the same time, the rich and diverse beauties of the Mission Beach and Cassowary Coast region have been threatened repeatedly by destructive mass tourism developments. These include the stealthy deforestation of ancient Gondwana-era lowland forests to generate new tourist access roads and to build environmentally destructive Gold Coast-style tourist amenities — processes that are wrecking the habitations and ecologies of the region's unique and most endangered species of fauna and flora. Such new and inappropriate tourist schemes are continually being planned or attempted. It seems to make no difference that recent bloated developer-driven enterprises have proved to be chimerical — not least because of the prodigious insurance costs that are now associated with building and maintaining expensive resorts in such a climate-sensitive region. Doubtless similar threats from mass tourism are familiar to many other communities that border the Great Barrier Reef and its adjacent rainforest. At a global level, they derive from enduring Western tussles between economy and ecology, with industrial capitalism having decided that nature matters only as an economic commodity. Ironically, being at the junction of two celebrated UNESCO-listed World Heritage Areas seems only to have exacerbated the Cassowary Coast's vulnerability to such inappropriate tourist-driven human interventions.

As a result, I have for some years been working with committed nature lovers and environmentalists in the region to find ways to contribute to its economic health without imperiling its rich biodiversity. Reading over this special issue has helped with this mission: both Kerrie and I noticed four themes that are explored in depth by our contributors, and which seemed pertinent to this task. These were: 'First Nations peoples', 'conservation heritages', 'creative arts traditions', and 'human–animal interactions'. All four, it struck us, had the potential to generate 'slow', 'knowledge-based' and sustainable forms of nature-culture tourism within the Mission Beach and Cassowary Coast region without endangering its biodiverse environments, wonders and ecosystems.

For this region possesses some superb, if largely untapped and under-used, nature-culture features. The three large Family Group islands of Dunk, Bedarra, and Timana adjacent to the towns of Mission Beach and Tully offer one such example. Thanks to their freehold land, ease of access to the mainland and sublime reef and rainforested environments, these islands have attracted a sustained history of nature arts movements that began with the arrival of 'Beachcomber' Ted Banfield in the late nineteenth century and continued up to the 2000s — since then soaring land costs have driven the artists to move to the mainland. The region's relatively numerous national parks and trails also continue to provide habitats for rare and charismatic wildlife populations, including the endangered Southern Cassowary and the Mahogany Glider. Furthermore, despite the relatively small surviving numbers of local First Nations Djiru people since their forced transfer to Palm Island in 1915, a group of talented and energetic individuals have nevertheless managed to revive the culture and the unique blend of forest and sea water traditions that shaped the nature/culture of this region for centuries.

Developing nature/culture tourism in the regions

Below I offer a few tentative specific suggestions for how the region's communities might develop potential forms of nature-culture tourism based on these four themes.

1. Creative arts tourism

Designed to celebrate the rich history and cultural heritage of the artists, craftspeople and nature writers who lived and worked on the Family Group Islands from the 1890s to 2000s, and whose legacies continue to flourish on the Mission Group mainland today.

The fascinating stories and cultural legacies of the region's still vibrant island arts and crafts movements cry out for appropriate forms of tourist-oriented recognition.

Their heritage encompasses the works of artists, craftworkers and nature writers such as Ted Banfield, Noel Wood, John Busst, Yvonne Cohen, Valerie Albiston, Bruce Arthur, Frederick Werther, Deanna Conti, Anneke Silva, Leonard Andy, Tom Risley, Helen Wiltshire, Peter and Helen Laycock, Liz Gallie and Susi Kirk — and this is by no means a comprehensive list of those artists who lived and worked for substantial periods of time on the islands since the 1890s. Moreover, others who visited for shorter periods to paint these dazzling tropical island landscapes include some of Australia's most famous names, such as John Olsen, Clifton Pugh and Fred Williams. The Family Island artists and craftspeople also established traditions that have persisted and evolved up until the present day. In the process, they pioneered a distinctive movement of modernist art that has helped to shape the wider aesthetic and lifestyle we now associate with Australia's Tropical North. From the time of Banfield onwards, many of the 'Beachcomber' Island artists and writers also became passionate communicators and protectors of Cassowary Coast flora and fauna.

Arts festivals, retrospective and contemporary art exhibitions, art island tours and Indigenous art trails are all examples of how this unique, enchanting nature-culture heritage could be celebrated and deployed to attract economic tourism without environmental destruction.[1]

2. Conservation tourism

Designed to experience the stories and sites of John and Alison Busst's heritage-listed Bingil Bay homestead, 'Ninney Rise', which functioned as the headquarters of heroic local conservationist campaigns from the 1960s to the 1980s to save and preserve Australia's world-famous Great Barrier Reef and Wet Tropics Rainforest areas.

Mission Beach boasts a lovely cyclone-proof bungalow and 0.8 hectare garden overlooking Bingil Bay that was designed and hand-built by artist-craftsman John Busst in the early 1970s and later bequeathed to the Queensland Parks and Wildlife Department (QPWD). Thanks to local community activists, it was subsequently awarded Queensland Heritage status in 2010 for its cultural, aesthetic and historical significance. In 2013, QPWD signed a long-term lease arrangement with the 'Friends of "Ninney Rise"' (FoNR), a non-for-profit organisation formed from a coalition of local community and environmental groups. FoNR's main objectives are to save the property from sale; to restore the building and grounds; and to attract local support to communicate the building's magnificent conservationist heritage as the headquarters of the local campaigns to save the Great Barrier Reef from oil, gas and fertiliser mining and the Wet Tropics rainforests from deforestation. Since 2013, FoNR has hosted fundraising open days and a series of successful public events, including seminars with Australian and international university science and humanities researchers. In December 2015, QPWD also granted FoNR long-term custodianship of another major Queensland Heritage listed site, Edward Banfield's Gravesite on Dunk Island.

FoNR has since commissioned a detailed Conservation Plan for 'Ninney Rise' and the John Busst Memorial in Bingil Bay that, after extended community consultations, listed exciting potential future aspirations for the site. These included attractive nature-culture tourist activities such as developing a small museum, library and archive; conducting school and public wildlife information workshops and tours; hosting university research groups and seminars; and offering small-scale residential programs, workshops and exhibitions for artists, scientists and writers. Recent assistance from QPWD and donors has also helped to restore the building and grounds so that 'Ninney Rise' can proceed in the near future to implement these exciting examples of culture-nature and heritage tourism.[2]

3. First Nations tourism

Designed to assist the key work of local Djiru artists and Traditional Owners to communicate and celebrate the historical, cultural, spiritual and environmental heritages of their people and Country.

International bodies, including UNESCO, have noted a mounting global interest in modes of slow, sustainable nature-culture tourism that entail respectful and sustained engagements with Indigenous peoples, to learn about their histories, cultures, spiritual values and complex interrelationships with local flora and fauna. Here too, the Mission Beach and Cassowary Coast region offers exciting opportunities to develop forms of Indigenous tourism that could help to nurture and sustain, rather than erode and exploit.

Despite their harsh colonial history of exploitation, the surviving mainland Djiru people of Mission Beach and Innisfail have, through the energy and talent of a small group of local activists, elders and artists, managed to initiate a renaissance of their history, culture, ecologies and art traditions. These also demonstrate how fundamentally the Djiru cultural heritage remains interlinked with the natural flora, fauna and geological features of their ancient Country. Archaeological, linguistic and oral history surveys have revealed scores of Djiru heritage sites within the Family Islands and on other areas of the Cassowary Coast mainland. Sites such as Clump Point, Boat Bay, South and North Mission Beach, Tam O'Shanter Point and Kennedy Bay are rich in ancient fish traps, ceremonial grounds, oven trees, stone formations, shell middens and stone artefact scatters.[3]

Prominent in fostering this renaissance is the distinguished artist and senior member of the Girringun Aboriginal Art Gallery in Cardwell, Leonard Andy. He lives on country and works tirelessly to reintroduce dispersed members of his clan to their former lands, culture and traditions. He and his daughter, Whitney Rassip, also disseminate the stories and culture of the Djiru people within the larger Cassowary Coast region by creating interpretative heritage signage that is featured in key heritage sites such as the Hull River Settlement Memorial at South Mission Beach and the Ulysses Heritage Trail at Mission Beach.[4]

In 2011, the Djiru people — past and present — were recognized as the traditional Native Title holders of their land in the Mission Beach area; two years later, Leonard and Whitney Rassip joined with several other traditional owners to procure title over two blocks at Wongaling Creek and South Mission Beach for which they have also drafted an integrated source management plan.[5] Leonard and Whitney have recently established an office in Mission Beach, from which they will continue to communicate and promote the multiple heritages, rights and achievements of the Djiru people.[6]

Having already developed a node to restore and expand forms of cultural and heritage knowledge for their own exiled peoples, the Djiru people of this region are thus well placed to respond to the growing national and international hunger of 'knowledge tourists' and visitors to learn about the unique Indigenous ecological knowledges and histories and artistic and spiritual traditions of this beautiful and fascinating region.

4. Human–animal tourism

Designed to communicate and expand community and visitor knowledge and understanding of the endangered Southern Cassowary, and to enhance means of protecting them from human threats.

Djiru Traditional Owner and artist Leonard Andy is also a respected local conservationist with a profound knowledge of the region's largest remaining community of endangered Southern Cassowaries, or *Gunduy*. With a wry smile, he points to his analogous involvement with the Cassowary species: '[It's] probably because I've lived with them a bit. They're going through what we went through ... with loss of habitat ... and life ... After the Cassowary Recovery Plan, when they were relocating them, it's a bit like their version of the Stolen Generation.' Moreover, he points out that Indigenous involvement in the care and plight of the Southern Cassowary has been formally mandated to Djiru people through having been signatories to the Wet Tropics Regional Agreement for the management of the Wet Tropics World Heritage Area. Leonard's Cassowary knowledge thus reflects a long mutual history of interaction between the Djiru people and Southern Cassowaries.[7]

The year 1991 also saw a sharp rise of non-Indigenous regional concern for the plight of the Cassowary through the foundation of several local community organisations that merged three years later to form the Community for Coastal and Cassowary Conservation. Commonly known as C4, its members have helped to instantiate the Cassowary as a symbol under which to unite the region's disparate community, government and environmental interests and concerns.

Even so, the need to preserve the Southern Cassowary has increased in urgency despite the success of the energetic president of Mission Beach Cassowaries, Liz Gallie, and her colleagues in organising major regional Cassowary Festivals to raise consciousness of and concern over the birds' plight. It is now imperative that we somehow find ways of halting the accelerating destruction of Cassowary habitats through urban spread, the reduction of their home ranges, intensified cyclone habitat damage and rising numbers of deaths from vehicle strikes and dog attacks. Above all, there remains an urgent need to investigate whether a sustainable form of Cassowary wildlife tourism is achievable. While tourist bodies have been quick to utilise the symbol of the Cassowary as a magnet for visitors to the region, they have been less successful in finding ways and means to ensure sustainable human–Cassowary interactions. The stakes could not be higher. Through their dispersion of wild fruit seeds, Cassowaries are also crucial to the survival of the region's lowland rainforests.

My warm thanks to all our contributors for inspiring these tentative reflections on how we humans might learn to develop some genuinely sustainable and benign forms of nature/culture tourism. As is evident in this special issue, change is – and simply must be – afoot in how we figure nature–culture relations, and this matters to the everyday lives of human and non-human communities such as those of Mission Beach. Our contributors capture the breadth of thinking necessary to create stories of the Great Barrier Reef and Wet Tropics Rainforests that seize the possibilities and opportunities which emerge when we have the courage to challenge and change the status quo.

Notes

1. See Ross Searle, *Artist in the tropics: 200 years of art in North Queensland* (Townsville: Perc Tucker Gallery, 1991; Ross Searle, *To the Islands: Exploring works created by artists in Dunk, Bedarra, and Timana Islands between the 1930s and 1990s* (Townsville: Gallery Services, 2013); Gavin Wilson, *Escape Artists: Modernists in the Tropics* (Cairns: Cairns Regional Gallery, 1998).
2. Michael Gunn and Catherine Brouwer, *Ninney Rise and John Memorial, Bingil Bay, North Qld Conservation Management Plan* (Cairns: Landscape Architects, 2016).
3. Pentecost, Indigenous Cultural Significance Assessment, pp. 14–32.
4. Leonard Andy, Girringun Aboriginal Art Centre, http://art.girringun.com.au/girringun-artists/leonard-andy, 18/9/20. 5.27pm
5. Djiru People Managing Country at Mission Beach, https://www.nrmq.org.au/djiru-people-managing-country-at-Mission-beach.
6. I am indebted to Constance Mackness, *Clump Point and District: An Historical Record* (Cairns: Bolton, 1983) for much of this information.
7. Pentecost, Indigenous Cultural Significance Assessment, pp. 44–6.

www.ingramcontent.com/pod-product-compliance
Lightning Source LLC
Chambersburg PA
CBHW060939170426
43195CB00022B/2980